RETHINKING INVASION ECOLOGIES FROM THE ENVIRONMENTAL HUMANITIES

Edited by Jodi Frawley and Iain McCalman

Routledge
Taylor & Francis Group
LONDON AND NEW YORK

earthscan
from Routledge

First published 2014
by Routledge
2 Park Square, Milton Park, Abingdon, Oxon, OX14 4RN

and by Routledge
711 Third Avenue, New York, NY 10017

Routledge is an imprint of the Taylor & Francis Group, an informa business

British Library Cataloguing in Publication Data
A catalogue record for this book is available from the British Library

Library of Congress Cataloging-in-Publication Data
Rethinking invasion ecologies from the environmental humanities / edited by Jodi Frawley and Iain McCalman.
     pages cm. – (Routledge environmental humanities)
   Includes bibliographical references and index.
   1. Biological invasions–Social aspects. 2. Introduced organisms–Social aspects. 3. Science and the humanities. I. Frawley, Jodi, author, editor of compilation. II. McCalman, Iain, autor, editor of compilation.
   QH353.R48 2014
   577'.18–dc23   2013033346

ISBN: 978-0-415-71656-7 (hbk)
ISBN: 978-0-415-71657-4 (pbk)
ISBN: 978-1-315-87964-2 (ebk)

Typeset in Bembo by
Keystroke, Station Road, Codsall, Wolverhampton

If Charles Elton's classic, *The Ecology of Invasions*, changed forever the way we thought about plants and animals, so too Frawley and McCalman's book is a major turning point. *Rethinking Invasion Ecologies* is a bold set of essays. Assimilation, migration, resilience, habitat, natives – all the conceptual ground that ecology, history and politics share is incisively explored. From crocodiles to humans, cane toads to prickly pears, in new worlds and old, this is environmental humanities at its sharpest.

Alison Bashford, University of Cambridge, UK

This exciting, timely and important collection illuminates the complex range of human values and actions that emerge from multi-disciplinary reflection. Often construed as relevant only to biologists, by adapting a cultural and historical perspective the innovative scholarship of *Rethinking Invasion Ecologies from the Environmental Humanities* reframes and reconceptualises the entangled, contradictory and ambiguous relationships between people and unruly biota.

Jane Carruthers, University of South Africa, South Africa

With pieces ranging from a biography of the concept of resilience to case studies of our reactions to cane toads, *Latrodectus* spiders, and salt-water crocodiles, this book argues the humanities have much to contribute to discussions of the Anthropocene. It makes a strong case, and its emphasis on Australia adds, for the rest of us, another voice to the dialogue. A fine collection on a fascinating and timely topic.

Thomas R. Dunlap, Texas A&M University, USA

This book demonstrates the value of the current turn to interdisciplinary approaches within a world transformed by colonial and postcolonial connections. Seeing human, plant and animal mobilities as thoroughly intertwined products of the Anthropocene, it innovatively bridges nature and culture and merges environmental, cultural and political histories.

Alan Lester, University of Sussex, UK

We know enough about the ecology of many invasive species to inform management that would make a huge difference. Around the world, the stumbling block is not so much the shortage of knowledge on what to do, it is ways of getting around the many complexities of the human dimensions of biological invasions. This book provides a crucial advance in this direction.

David M. Richardson, Stellenbosch University, South Africa

CONTENTS

FIGURES

CONTRIBUTORS

Christina Alt is a Lecturer in the School of English at the University of St Andrews, Scotland. Her research focuses on historical intersections between literature and science. Her research interests include the discourse surrounding ecology as it emerged as a scientific discipline in the early twentieth century and the impact of this discourse on modernist representations of nature.

Brett M. Bennett is a Lecturer in Modern History at the University of Western Sydney Australia. He is pursuing research projects on the global history of the world's forests. He is developing ways to use history as an applied social science and scientific methodology to improve development and environmental management decision-making, particularly in regions surrounding the Indian Ocean.

Gilbert Caluya is an Australian Research Council DECRA Fellow and Research SA Fellow of the MnM Centre at the University of South Australia. He has published widely on race and sexuality and has worked on two funded projects for the Defence Science and Technology Organisation. His current research explores the relationships between race, intimacy and security.

Andreas Aagaard Christensen has an MSc in Geography and Cultural Studies, and a BA in Psychology from Roskilde University, Denmark. He is a PhD Fellow at the Department of Geosciences and Natural Resource Management, Copenhagen University, Denmark. His research project is the study of land use culture among farmers in European and colonial agrarian landscapes and the effect of such subjectivities on the ecology of rural landscapes.

Jodi Frawley is a Discovery Early Career Researcher Award Fellow in the Department of Gender and Cultural Studies, University of Sydney, Australia. She is interested in the transnational networks that formed through the global movement of plants and their consequent impacts on local environments, especially those species named as invaders.

Lesley Head is an ARC Australian Laureate Fellow and Director of the Australian Centre for Cultural Environmental Research at the University of Wollongong, Australia. She has researched human–plant relations in diverse contexts including Aboriginal land management, backyard gardens, wheat farms and invasive plants.

Peter Hobbins is a historian of science and medicine. His research to date has focused on the structures and projects of Australian natural history and biomedicine to 1945. He recently completed his PhD on venomous animals in colonial Australasia and is now a Research Associate within a conjoint history-archaeology project on quarantine at the University of Sydney, Australia.

Christian Kull teaches at the School of Geography and Environmental Science, Monash University, Australia. His research extends across political ecology and development geography, with particular interest in fire ecology and natural resource management in Madagascar, South Africa, and northern Australia.

Peter Marks teaches in the Department of English at the University of Sydney, Australia. He has published on surveillance and utopia, film, social realism in the 1930s, literary periodicals and Margaret Atwood.

Iain McCalman is a Professorial Research Fellow in History, University of Sydney. He is Co-Director of the Sydney University Environment Institute, Australia. His research areas include maritime and environmental history of the sea and the uses of multimedia for historical storytelling in the environmental humanities.

Cameron Muir is a Postdoctoral Fellow at the Australian National University and Fellow at the Rachel Carson Center of Environment and Society, LMU, Munich. His main research interest is in relationships between food, agriculture, environment and social justice. He is currently working on projects examining the origins of the current food crisis, and the social and environmental changes wrought by agricultural pesticides.

Emily O'Gorman is an environmental historian with interdisciplinary research interests. Her research within the environmental humanities focuses on how people live with rivers, wetlands, and climates. Currently a Lecturer at Macquarie University, Australia, she holds a PhD from ANU and undertook a postdoctoral candidacy at the University of Wollongong, Australia.

Eric Pawson is Professor of Geography at the University of Canterbury, New Zealand. His research interests span environmental history, landscape development and the futures of New Zealand's biological economies. He was awarded the Distinguished New Zealand Geographer Medal in 2007.

Simon Pooley is an environmental historian embedded with Imperial College Conservation Science, London. He aims to articulate the role of historical research in the context of nature conservation, discover ways of improving multiple disciplinary collaborations in research, and contribute impactful research on the conservation of large predators. His case study is international crocodilian conservation post 1945.

Haripriya Rangan teaches at the School of Geography and Environmental Science, Monash University in Melbourne, Australia. Her research falls within the ambit of political ecology and environmental history, with particular interest in regional economic change, development, and natural resource management in South Africa, India, and northern Australia.

Morgan Richards is interested in the interplay between animals, new media technology, environmental politics and public service media. She is currently researching the Australian wildlife television industry for a new collaborative project with Professor Gay Hawkins and partner investigators at the ABC called 'Making Animals Public'.

Harriet Ritvo is the Arthur J. Connor Professor of History at MIT, USA. She has research interests in British history, environmental history, the history of human–animal relations, and the history of natural history.

Libby Robin is an environmental historian at the Australian National University and the National Museum of Australia. She is also Guest Professor at the Royal Institute of Technology (KTH), Stockholm, working in the KTH Environmental Humanities Laboratory.

Anna Wilson obtained her doctoral degree from the University of Tasmania, Australia. She is currently exploring connections between gardening and plant naturalisation. She worked with Kull and Rangan on her Master's research at Monash University, Australia. Her Master's thesis, entitled 'Doing right by country: the pastoral industry and prickly bush management in northwest Queensland', provided the empirical basis for their chapter.

ACKNOWLEDGEMENTS

This book started life as a two-day symposium at the University of Sydney in July 2012. Over 50 participants and invited speakers came together on those days to grapple with the ideas that humanist scholars have produced in response to the scientific research on invasive species and their impacts in a range of environments across the world. We thank all the participants and speakers for such an exciting exchange of ideas. Our contention that scholars from across the humanities have something of value to add to debates about how we might cope with environments changed by the geological forces of the Anthropocene gathered pace over these two days, and the months that followed the conference.

We thank the Environmental Humanities Research group that provided the initial ideas for the conference and the funds made available by the Faculty of Arts Collaborative Research Scheme. Alison Bashford deserves special thanks for her unstinting support and encouragement for the project. David Schlosberg, Co-Director of the Sydney Environment Institute and the Sydney Network for Climate Change and Society enthusiastically contributed to our establishment, and has been instrumental in encouraging the development of scholarship on the environment across all the faculties of the University of Sydney.

Elspeth Probyn, Peter Hobbins, Lynette Finch and Kate Fullagar all gave different kinds of valuable advice along the way. We thank Katherine Anderson and Michelle St Anne for their patience, dedication and organisational skills that helped to shape this book.

Chapter 13, 'Naturalising Australian Trees in South Africa' is reprinted from the *Journal of Southern African Studies.*

Finally, our thanks to Libby Robin, Khanam Virjee and Helen Bell from Routledge for their seamless editorial advice and enthusiasm for the collected work of all the authors in this volume.

PART I

Setting the scene

1

INVASION ECOLOGIES

The nature/culture challenge

Jodi Frawley and Iain McCalman

In 1984, the American comedy/horror film *Gremlins* hit the screens across the world. A little furry animal, Mogwai, is given to Billy Peltzer as a pet with instructions to keep it away from light, not to let it get wet and never to feed it after midnight. But Billy inadvertently spills water on the creature, causing it to multiply, and the beguiling animal also tricks its human companions into a late-night feeding. Instantly the cute fur-ball is transformed into a reptilian gremlin whose rapid reproduction changes the scenario from controlled domesticity to explosive over-population. The proliferating gremlins then run amok in Billy's hometown, altering the landscape to suit their own needs. The town swimming pool, for example, becomes a fertility site, its waters seen by the creatures as a perfect breeding place to accelerate their expansion. Gremlins proceed to frighten the residents, destroy the shops and go berserk through the schools. In short, they become a threatening invasive species, which has to be dealt with by Billy and the town folk.

All the key elements of invasion are present in this story. There is the link with humans who willfully move this biological species out of one habitat into their own. A new environment is provided for the animal, where it appears to be right at home, so long as it is contained within human rules. Water and food are environmental cues that trigger its multiplication. Invasion is shaped by the fecundity of the species. All of these things move rapidly out of synchrony with the host community. And as the conditions of life alter this biological entity, its interactions with humans change too. Relations between species shift from congenial to fearful as the animal's transformation into an invader becomes evident. Where it had once been a companion, it was now regarded as dangerous matter out-of-place. Christopher Smout sums this up in saying 'to be an alien is not a biological characteristic, like being blue or having a square stem; it is a character implied by man' (Smout, 2011).

In other words, it is the way that species interact within bio-cultural environments, rather than their individual biological characteristics, that results in the formation of invasion ecologies (Hobbs *et al.*, 2006). These formations are not restricted, however, to isolated local places. They also include multilayered geographies. A series of sites, identified by Haripriya Rangan and Christian Kull (2009a), are connected by the global movement of the acacia species, camels and cameleers around the Indian Ocean. In this example, these plants and animals transferred between India, South Africa and Australia. They moved along with bundles of knowledge that constantly remade places at a range of geographic scales: local, regional, national and global (Kull *et al.*, 2007; Kull and Rangan, 2008; Rangan and Kull, 2009b). These conditions are always understood in relation to human cultural discourse—whether they emerge from science, management, history, economics, or the environments themselves. Each configuration is as specific as it is complex.

Not surprisingly, one influential account of the origin of the Anthropocene—the first ever human-shaped geological era—has it begin in 1784 with the development and use of the steam engine (Robin and Steffan, 2007: 1699). Fundamental atmospheric change in the form of increased greenhouse gases and global warming are viewed as an unintended outcome of this revolution in human ingenuity. Another of the human-driven changes of the Anthropocene, however, arises from the rampant overabundance of introduced species around the globe, a diaspora of nature resulting, in some cases, in the crippling of the new country's ecological health and balance. The results of this process have become generally known as invasion ecologies (Mooney, 2005; Mooney *et al.*, 2005; Richardson, 2011). Research from the sciences and social sciences has been at the forefront of refining both the concept of an Anthropocene era and the emerging community understandings about what living with its consequences will mean for the future (McNeely, 2001; Rotherham *et al.*, 2011).

The chapters in this book build on this achievement by demonstrating how research derived from a humanities perspectives can transform our understandings of the character and implications of invasion ecologies (Hall, 2003; Robbins, 2004a; Head *et al.*, 2005). Furthermore our contributors are in agreement that modern environmental approaches that treat nature with naïve realism or as a moral absolute, unaware or unwilling to accept its entanglement in cultural and temporal values, are doomed to fail. We need rather to investigate the complex interactions of ecologies, cultures and societies in the past, present and future if we are to understand and solve the current problems of the global environmental crisis (Comaroff and Comaroff, 2000; Parker, 2001; van Dooren, 2011; Rose *et al.*, 2012). During the Anthropocene time frame, environments of the new world—over both land and sea — have become testing grounds for the introduction of new assemblages of people and plants, economies and animals, cultures and coastlines (Beinart and Middleton, 2004, Johnson, 2010). The resultant environmental changes often led to unexpected and enduring ecological and social impacts, some adverse, some beneficial: impacts that we now know to be dynamic, unpredictable and often contradictory. Above all, our authors interrogate the complex and ongoing community concerns about invasive species and their ecological and cultural impacts that we will together have to face in a climate-changing world.

At the same time our book does not pretend to survey the invasion ecologies field in all its facets but, rather, to explore a linked set of themes that have become particularly salient in our time. How, we ask, will biological and cultural invasions of the past influence the present and futures of climate-changing places? How should we think about the more-than-human roles of camels and carp, or of willows and baobabs? What became of the plants, animals, people and ideas that traveled and re-made other countries and places in the pursuit of empires? From the late eighteenth century onwards the New World countries of Australia, New Zealand, Africa and the Americas all became laboratories for western science and colonization (Fullagar, 2012). As a result, these postcolonial places furnish especially rich examples of where the movement of biota has disrupted, degraded and altered ecosystemic relationships across the globe.

Another of our aims has been to multiply disciplinary conversations within the burgeoning field of the environmental humanities in order to explore how conceptual understandings of invasion ecologies can be infused with a variety of literary and artistic narratives, gendered tropes and moral fables, as well as with more familiar political, legal, and sociological inflections. Human beings, cultures and natures have been, and remain, deeply entangled and interdependent in these new landscapes of empire and post-empire in ways that only the environmental humanities can uncover.

Charles Elton's (1958) classic study, *The Ecology of Invasions by Plants and Animals*, signaled a major shift in the understanding of the global movement of biological species during what we now think of as the Anthropocene (Elton, [1958] 2000; Richardson, 2011). Over the nineteenth and twentieth centuries, new plants, animals and humans migrated to settler colonies at the same time that biological materials and ideas about nature were transiting to other parts of the world. Some of these migrating species became threats to local environments wherever they lodged. Famously framed by Alfred Crosby as 'the Columbian exchange' (1972) and 'ecological imperialism' (1986), this tidal wave of people and their non-human accompaniments surged across the globe as empires came and went during the period from the sixteenth to the twentieth centuries. By the 1950s, however, the resulting acclimatization and naturalization movements were being challenged by new scientific attempts to manage the unforeseen ramifications of changes to landscapes, environments and ecologies (i.e. the study of relationships between living organisms and their environments).

Scientific-based approaches to invasive species were consolidated in the 1980s with the advent of invasion biology and the related disciplinary fields of conservation biology and restoration ecology (Simberloff and Rejmánek, 2011). These new disciplines focused dispassionately on the non-human world to analyze relations within natural ecosystems on both land and sea (Mooney *et al.*, 2005). Yet wherever humans had remade local places in ways that compromised functionality, complex ethical and normative issues arose that many scientists felt to be outside of their remit. Other social activists and thinkers, believing that humans had incurred a responsibility to protect both present and future environments, urged the need for fresh interventions in order to restore lost places and optimal outcomes. It was in this context and in an effort to create different, more sustainable futures that a pressing need developed for humanities research to contribute to social and cultural interpretations of invasion ecologies.

Admittedly, the disciplines of history and geography have long had some limited acceptance within the primarily scientific-based fields that make up modern invasive ecology studies (Kitching, 2011; Henderson et al., 2006; Carlton, 2011). Historians have sometimes been co-opted to assist in the discovery and development of the temporal and spatial baselines that are so critical to the work of sciences that investigate organic invasions (Brown et al., 2008). Ecologists draw data from historical sources, which are then fed into computational modeling for restoration work (Bjorkman, 2010). Without such data, many of the kind of models commonly in use in conservation and restoration work would be meaningless (Anderson, 2006; Bolster, 2008). Environmental geographers too have been encouraged to contribute in specific ways to invasion ecology research (Robbins, 2001, 2004b; Pawson, 2008). By creating different scales for ecological analysis, biogeography has been particularly important in mapping the ways that species move into new places (Davies and Watson, 2007).

Yet there is a danger that historical and geographical research of this kind is viewed merely as a handmaiden to invasion ecology work driven almost wholly from the perspectives of the natural sciences. By expanding the range of humanities scholars involved in this field of study, our hope is that new questions will emerge to complement and challenge those driven solely by science, important though these are. Here, such fields as eco-criticism, cultural geography, indigenous studies and environmental philosophy can offer both complementary and alternative ways of analysis, interpretation and understanding.

One key such contribution is the mapping of 'shifting baselines'. This refers to the way that each generation, ignoring prior historical conditions, blindly considers their own ecological circumstances to be the foundation for all decision-making in science and policy. Understanding the impacts of anthropogenic change requires, however, that historical evidence and contemporary cultural theories be incorporated into policies, regulations and popular understandings (Jackson et al., 2011). Although there are many ways to assess such vital benchmarks, most have up to now been dominated by scientific classifications of 'nativeness', and by value-laden notions of what constitutes stable, balanced or healthy ecosystems. Restoration work, conversely, assumes that ecologies have been corrupted, put under threat or thrown out of order. Here ecologists tend to conjure up perfect ahistorical pasts, which contemporary scientists, managers, and communities then work to recapture (Alagona et al., 2012). All too often nostalgia has subsumed a reality that is much more ambiguous. Research grounded in the experiences of people engaged in grappling with changed or changing environments often proves more complex, revealing that new ecosystems may be detrimental to some actors, while others can be energized by these same disrupted, hybrid or changing environments (Davis *et al.*, 2011).

Part II Invasion and the Anthropocene

Few would dispute that the environmental problems we face today link directly to the Age of Empires from the eighteenth century through to the present. In Chapter 2, distinguished historian Harriet Ritvo explores the back story to global ecological change by investigating the imperial movements of biological species. Colonial acclimatization movements that were set up to facilitate transfers of useful or economic plants and animals into new worlds provided much of the ideological and infrastructural momentum for such change. Ritvo focuses on starlings and camels as subjects of acclimatization, each with their own histories of global travel. Acclimatization stories of this kind can be multiplied for species and nations, across and between empires and in oceans and soils of the earth. Here we see one compelling example of how *anthropoi* animated the new geological era of the Anthropocene.

The concept of the Anthropocene intersects with invasion ecologies in a variety of other ways as well. In Part II, we also situate invasion ecology within wider historical and scientific frameworks in order to explore the broader narratives and languages of nature and scientific work that the field encompasses, and which have helped to shape its history and shadow its future. We range from the vast geological and historical time scales of the new age of the Anthropocene, to the more recent and widely invoked scientific concept of resilience, to the microcosmic landscape changes that can be traced in particular local sites.

Racial formations in a globalized world, for example, are shown to resonate throughout the metaphors and allegories of invasion. Gilbert Caluya's Chapter 3 presents lively provocations from a postcolonial perspective to the ways that the concept of the human has been deployed in recent articulations of the Anthropocene, a concern that he extends even to Chicago historian Dipesh Chakrabarty's essay, 'The Climate of History: Four Theses', published in 2009 in *Critical Inquiry*. Despite its great synthesizing power, Caluya warns, the concept of a unified humanity implicit in the definition of the Anthropocene risks returning us to a universal metanarrative that sweeps aside the sharp distinctions that continue to operate within the structures of race and class under advanced capitalism.

Historian of science and environment Libby Robin analyzes the pervasive scientific concept of 'resilience' in relation to invasion biology in Chapter 4. Tracing the historical biography of 'resilience', Robin shows how the concept emerged, was adopted and extended by influential scientists, and was ultimately shaped through the interplay of experiences within a range of arid and desert ecosystems. In the process she shows how scientists have oscillated between global uses of the term and applications derived from intimate local understandings of place. Resilience, Robin predicts, will remain a pivotal idea for ecological invasion study as long as the environmental conditions in the Anthropocene continue to present urgent challenges for humanity.

Where Caluya and Robin focus mainly on global applications of the Anthropocene concept, in Chapter 5 geographers Eric Pawson and Andreas Aagaard Christensen illuminate the histories of a fascinating local site, the Banks Peninsula on the South Island of New Zealand. Here they trace the changing interrelations between the land and its flora with the temporal layers of human occupation, demonstrating the ways that valued introduced species of grasses and other fodder plants actively displaced indigenous forests. To the local Maori clans the materiality of these newly remade places was experienced as a form of colonization that displaced their clans in tandem with the trees. Yet Pawson and Christensen also show that what was celebrated by generations in one period could be vilified as invasive in the next. Concluding with an examination of the biotic formations on the Banks Peninsula today, they suggest that in the future the locality is likely to be dominated by hybrid environments where a patchwork of meanings compete within the same spatial zones.

Part III Everyday life in invasion ecologies

Managing proliferating weeds, pests, invaders and aliens has often evoked the hyperbole of crisis and disaster. However, when responding to the challenges of climate change on more local scales, humans are likely in future to have to adapt their environments and behaviors without reference to idealized images of past ecologies. In particular, we may need to think more flexibly about the cultural values entailed in supposed ecological invasions. Removing whole populations of invasive species is unlikely to prove as viable and defensible as it has in the past. Hard questions of ethics will generate robust debate, in which humanities scholars will undoubtedly feature. At the same time we must be prepared to scrutinize our own practices and work to generate collaborations that reach across the disciplinary divides.

Once animals and plants are perceived to move from naturalized or stable populations to become invasive, such a classification demands that they be managed. Since the nineteenth century the multiplicity of different kinds of invasion have produced an array of different techniques and technologies of intervention, eradication and restoration, including experiments with poisons, biological controls, human labor and machines.

Practices and technologies like these must of course operate within particular rules, regulations, and legal systems that set parameters designed to bring invading biota under control. Our contributors also interrogate a range of such everyday problems and potential solutions. How might garden weeds help us to think through what is needed for the future? What new plant management ideas can emerge from situations where people reconcile themselves both to living with change and to experimenting within degraded environments? How have colonized peoples responded over the years to the many new species that have entered and reshaped their landscapes?

In Chapter 6, cultural geographer Lesley Head uses garden weeds to think freshly about invasion ecologies. If we cannot live without weeds, the garden is one space where we can understand what it means to live with them. In their troublesome encounters, people experience the agency of plants. Weed management requires considerable investment of labor and vigilance. The most successful and contented garden weed managers accept that plants have a life of their own and do their own thing. Above all Head urges the need for us to relinquish nostalgic ideals of pristine pasts and accept that we face contradictory, uncertain, and weedy futures.

By investigating the work of an experimental station in western New South Wales, in Chapter 7, historian Cameron Muir brings a very different perspective to the invasion ecologies question. Rather than focusing on a specific species, which here might be wheat and the environmental degradation it often produces, he demonstrates the ways that scientific thought and experimental work operated within a discursive framework that ran counter to that of the Europeans who lived in and worked the region. Without shying away from the intergenerational destruction caused by the clearing of native vegetation in the pastoral economy, Muir complicates the standard picture by telling the story of Robert Peacock, whose deeply practical experience shaped the kinds of experiments undertaken at Coolabah.

In Chapter 8, geographers Haripriya Rangan, Anna Wilson and Christian Kull confront the complicated story of another invader—the introduction of prickly acacia in northern Queensland. Beginning with an outline of the 'native versus alien' debate within Queensland, they show how national narratives of land changed with the emergence of the new programmes, 'Landcare' and 'Caring for Our Country'. They explore these ideological developments within the context of the everyday experiences of the people who have to live within the altered landscapes, demonstrating that responses to invasive species are formed fluidly within a complex nexus of policy, larger economic forces, and lived practices.

Part IV Ecological politics of imagining otherwise

When a plant or an animal is identified as 'out-of-place', humans invariably generate an array of ways to imagine the intruders differently. These imaginings can range from dreams of landscapes without invaders to nightmares of escalating catastrophe. Imagined worlds of this kind tend to embody the anxieties, hopes and fears of different generations about invasive species, ecologies, landscapes and peoples. In Part IV, we examine how a series of invasion ecologies are imagined to be 'otherwise'. Although such imaginings are most evident within speculative literature and film, where strange and different futures and pasts can be explored alongside critiques of the human condition, we also explore how scientists and policy-makers similarly employ imagined futures to articulate invasion ecologies in order to promote their scientific research and communicate with their publics.

Christina Alt takes us into the realms of early science fiction with her Chapter 9, on the cross-over between late nineteenth-century scientific knowledge about invasive species and their simultaneous literary imaginings. Alt shows how the red weed evoked in H. G. Wells's (1898) *War of the Worlds* drew on contemporary knowledge about South American prickly pear *Opuntia* spp and Canadian waterweed *Elodea canadensis*. While Wells's machines and technologies carry one message about science, Alt argues, the multispecies environment of invasion proved equally important to the enduring social impact of the story.

In Chapter 10, Morgan Richards eloquently confronts us with the cultural impact of an equally bizarre real-life invader in order to demonstrate how an innovative use of documentary filmmaking can present a range of counter-hegemonic ideas about invasiveness and invasive species. Here she focuses on Mark Lewis's celebrated and wryly comical pair of films about the remorseless march through Australia of the species of poisonous cane toads, *Bufo marinus*, originally introduced from South America to control the north Australian cane beetle. Although cane toads are popularly reviled and spectacularly slaughtered, Richards shows that they have also evolved into culturally ambiguous creatures that shape the imaginaries of the humans around them, as much as they disrupt actual material ecologies.

In Chapter 11, Peter Marks takes us into the haunting dystopian environments of Margaret Atwood's *Oryx and Crake*, inhabited by cyborg people and biomanufactured animals that would make the cane toad seem cuddly. Atwood, Marks argues, raises concerns about invasive species in the age of the Anthropocene that have potential futures as well as identifiable pasts. The notion that new hybrids of humans and animals might emerge from future biotechnological processes crystallizes a sharp contemporary cultural contradiction: the idea that the technological/scientific matrix can solve all the toxic problems that it has itself created. Ecological invasion is invariably shaped by biological, ecological and environmental forces that are always subjected to the human desire for mastery. Yet, as Bruno Latour (2000) has observed, the unruliness of biological life continually surprises in our more-than-human worlds.

Part V Unruly natives and exotics

'Native', 'indigenous', 'exotic' and 'foreign' are concepts that have been mobilized all over the world in debates in order to ground both the science and the management of invasive species (Subramaniam, 2001; Chew and Hamilton, 2011). Rhetoric about species assists both community members and scientists to frame responses to invasion (Frawley, 2007; Keulartz and Weele, 2008; Smout, 2011). Clear demarcations between those who belong and those who invade have shaped eradication programs, policies and legislation, as well as the ways that human communities respond to changed environments (Smout, 2003; Tigger *et al.*, 2008, Lavau, 2011). However, the use of such stark binaries denies the dynamism of ever changing environments (Head and Muir, 2004; Beinart and Wotshela, 2012; Lennox *et al.*, 2012). Natives can and do become invaders in the very same ways that exotics can become overabundant. Conversely exotics can also be invasive in foreign places while being endangered in their native habitats (Frawley and Goodall, 2013).

The chapters in Part V challenge us to think more closely about the way that plants, animals and humans interact with one another and with different kinds of environments. Ritvo's Chapter 2 traced the historical origins of acclimatization, a mode of human activity that relied on the movement of biological material to reshape new places under Empire. Here, too, histories of crocodiles, ducks, spiders and trees demonstrate that close attention to local cases can disrupt both the native and exotic categories so often used within the rhetoric of invasion ecologies.

Our contributors show the value of opening up our thinking to more-than-human approaches (Whatmore, 2002; Probyn, 2011) and beyond these to the challenging perspectives of the more-than-animal and the more-than-plant as well. Thinking about natives and exotics also challenges us to encompass a range of scales since changing impacts on local peoples and species can carry through to regions, nations and ultimately to global communities.

In Chapter 12, Peter Hobbins tackles the ontological quandary of how invasive species become visible to the communities they inhabit by exploring how *Latrodectus* spiders in Australia and New Zealand acquired their venomous reputations. After tracking through archival records of redback spiders in Australia and Katipo spiders in New Zealand, he conjures up a revisionist history of migrant arthropods. In the process he reveals that getting to know spiders through the sites of painful contact has much to tell us about human interrelations with non-charismatic animals in colonial environments, as well as about the tenuousness of the creatures' strange ontologies.

Though less confronting than venomous spiders, Australian trees of the genera *Acacia*, *Casuarina*, *Eucalyptus* and *Hakea* that have found their way to South Africa are there both praised for their contribution to domestic economies and vilified as exotic invaders. In Chapter 13, Brett M. Bennett offers a detailed account of how the trees originally migrated to South Africa through governmental exchanges of knowledge, seeds, seedlings and plants. Forestry departments in South Africa used modern techniques of silviculture to match regional microclimates between the two countries in order to enhance the capacity for these particular tree species to survive and flourish in their new habitats.

The Murrumbidgee Irrigation Area of the Murray-Darling Basin in Australia represents a particularly charged site for debates about invasive species because changing government water regulations have caused the issue to become entangled in acrimonious environmental disputes over irrigation rights. Emily O'Gorman's Chapter 14 contextualizes this political contest by analyzing the persistence of ideas about the invasiveness of ducks and other water birds within the agricultural ecologies of twentieth-century rice growing when biologists James Kinghorn and Harry Frith were enlisted to study ducks at different times. She shows how particular duck species were categorized as invasive by scientists enlisted to defend grower interests, even when the birds' destructive properties were disputed by dissenting scientists and rice farmers.

If ducks are transformed into alien and unwanted species when they eat rice seedlings, crocodiles become instantly invasive when they eat people. Yet, as Simon Pooley wittily shows in Chapter 15, their reputation in Australia has been surprisingly ambiguous. Among some of the inhabitants of Northern Australia they are seen as charismatic and ancient creatures to be celebrated as emblems of local nativity, as tourist attractions in parks, and as skin providers for the shoe and handbag trade. As with venomous spiders, however, any substantial increase in their numbers gives rise to invasive panics based on their threat to visitors and perceived interference with local recreations. Though few humans will admit to loving the Australian 'Saltie', Pooley closes his chapter and our book with a plea that we take responsibility for the environmental changes of the Anthropocene that now are bringing all humans and all non-humans into inescapable and dynamic contact.

References

Alagona, P. S., Sandlos, J. and Wiersma, Y. F. 2012. Past Imperfect: Using Historical Ecology and Baseline Data for Contemporary Conservation and Restoration Projects. *Environmental Philosophy*, 9, 49–70.

Anderson, K. 2006. Does History Count? *Endeavour*, 30, 150–155.

Beinart, W. and Middleton, K. 2004. Plant Transfers in Historical Perspective: A Review Article. *Environment and History*, 10, 3–29.

Beinart, W. and Wotshela, L. 2012. *Prickly Pear: The Social History of a Plant in the Eastern Cape*, Johannesburg, Wits University Press.

Bjorkman, A. D. 2010. Defining Historical Baselines for Conservation: Ecological Changes Since European Settlement on Vancouver Island, Canada. *Conservation Biology*, 24, 1559–1568.

Bolster, J. 2008. Putting the Ocean into Atlantic History: Maritime Communities and Marine Ecology in the Northwest Atlantic. *The American Historical Review*, 113, 19–47.

Brown, S., Dovers, S., Frawley, J., Gaynor, A., Goodall, H., Karskens, G. and Mullins, S. 2008. Can Environmental History Save the World? *History Australia*, 5.

Carlton, J. T. 2011. The Inviolate Sea? Charles Elton and Biological Invasions in the World's Oceans. In: Richardson, D. M. (ed.) *Fifty Years of Invasion Ecology: The Legacy of Charles Elton*, Oxford, Blackwell Publishing.

Chakrabarty, D. 2009. The Climate of History: Four Theses. *Critical Inquiry*, 35, 197–222.

Chew, M. K. and Hamilton, A. L. 2011. The Rise and Fall of Biotic Nativeness: A Historical Perspective. In: Richardson, D. M. (ed.) *Fifty Years of Invasion Ecology: The Legacy of Charles Elton*, Oxford, Blackwell Publishing.

Comaroff, J. and Comaroff, J. L. 2000. Naturing the Nation: Aliens, Apocalypse, and the Postcolonial State. *HAGAR International Social Science Review*, 1, 7–40.

Crosby, A. W. 1972. *The Columbian Exchange: Biological and Cultural Consequences of 1492*, Westport, CT, Greenwood Publishing Co.

Crosby, A. W. 1986. *Ecological Imperialism: The Biological Expansion of Europe*, Cambridge, Cambridge University Press.

Davies, A. and Watson, F. 2007. Understanding the Changing Value of Natural Resources. *Journal of Biogeography*, 34, 1777–1791.

Davis, M. A., Carroll, S. P., Thompson, K., Pickett, S. T. A., Stromberg, J. C., Del Tredici, P., *et al.* 2011. Don't Judge Species on Their Origins. *Nature*, 474, 153–154.

van Dooren, T. 2011. Invasive Species in Penguin Worlds: An Ethical Taxonomy of Killing for Conservation. *Conservation and Society*, 9, 286–298.

Elton, C. S. [1958] 2000. *The Ecology of Invasions by Animals and Plants*, Chicago, University of Chicago Press.

Frawley, J. 2007. Prickly Pear Land: Transnational Networks in Settler Australia. *Australian Historical Studies*, 38, 323–339.

Frawley, J. and Goodall, H. 2013. Transforming Saltbush: Science, Mobility and Metaphor in the Remaking of Intercolonial Worlds. *Conservation and Society*, 11, 176–186.

Fullagar, K. (ed.) 2012. *The Atlantic World in the Antipodes: Effects and Transformations since the Eighteenth Century*, Newcastle: Cambridge Scholarly Publishing.

Hall, M. 2003. Editorial: The Native, Naturalised and Exotic: Plants and Animals in Human History. *Landscape Research*, 28, 5–9.

Head, L. and Muir, P. 2004. Nativeness, Invasiveness, and Nation in Australian Plants. *The Geographical Review*, 94, 199–218.

Head, L., Trigger, D. and Mulcock, J. 2005. Culture as Concept and Influence in Environmental Research and Management. *Conservation and Society*, 3, 251–264.

Henderson, S., Dawson, T. P. and Whitaker, R. J. 2006. Progress in Invasive Plants Research. *Progress in Physical Geography*, 30, 25–46.

Hobbs, R. J., Lugo, A. E., Norton, D., Ojima, D., Richardson, D. M., Sanderson, E. W., et al.. 2006. Novel Ecosystems: Theoretical and Management Aspects of the New Ecological World Order. *Global Ecology and Biogeography*, 15, 1–7.

Jackson, J. B. C., Alexander, K. E. and Sala, E. (eds) 2011. *Shifting Baselines: The Past and Future of Ocean Fisheries*, Washington, DC: Island Press.

Johnson, S. (ed.) 2010. *Bioinvaders*, Cambridge: The White Horse Press.

Keulartz, J. and Weele, C. V. D. 2008. Framing and Reframing in Invasion Biology. *Configurations*, 16, 93–115.

Kitching, R. L. 2011. A World of Thought: 'The Ecology of Invasions by Plants and Animals' and Charles Elton's Life Work. In: Richardson, D. M. (ed.) *Fifty Years of Invasion Ecology: The Legacy of Charles Elton*, Oxford, Blackwell Publishing.

Kull, C. and Rangan, H. 2008. Acacia Exchanges: Wattles, Thorn Trees, and the Study of Plant Movements. *Geoforum*, 39, 1258–1272.

Kull, C., Rangan, H. and Tassin, J. 2007. Multifunctional, Scrubby and Invasive Forests? Wattles in the Highlands of Madagascar. *Mountain Research and Development*, 27, 224–231.

Latour, B. 2000. When Things Strike Back. *British Journal of Sociology*, 51, 107–123.

Lavau, S. 2011. The Nature/s of Belonging: Performing an Authentic Australian River. *Ethnos*, 76, 41–64.

Lennox, S., Mulaudzi, R., Potgieter, M. and Erasmus, L. 2012. Not All Invasives Are Equal: Communities in Limpopo on the Prickly Pear. *Veld and Flora*, 70.

McNeely, J. A. (ed.) 2001. *The Great Reshuffling: Human Dimension of Invasive Alien Species*, Gland, Switzerland: International Union for Conservation of Nature and Natural Resources.

Mooney, H. A. 2005. Invasive Alien Species: The Nature of the Problem. In: Mooney, H. A., Mack, R. N., McNeely, J. A., Neville, L. E., Schei, P. J. and Waage, J. K. (eds) *Invasive Alien Species: A New Synthesis*, Washington, DC: Island Press.

Mooney, H. A., Mack, R. N., McNeely, J. A., Neville, L. E., Schei, P. J. and Waage, J. K. (eds) 2005. *Invasive Alien Species: A New Synthesis*, Washington, DC, Island Press.

Parker, V. 2001. Listening to the Earth: A Call for Protection and Restoration of Habitats. In: McNeely, J. A. (ed.) *The Great Reshuffling: Human Dimension of Invasive Alien Species*, Gland, Switzerland: International Union for Conservation of Nature and Natural Resources.

Pawson, E. 2008. Plants, Mobilities and Landscapes: Environmental Histories of Botanical Exchange. *Geography Compass*, 2, 1464–1477.

Probyn, E. 2011. Swimming with Tuna: Human–Ocean Entanglements. *Australian Humanities Review*, 51, 97–114.

Rangan, H. and Kull, C. 2009a. The Indian Ocean and the Making of Outback Australia: An Ecocultural Odyssey. In: Moorthy, S. and Jamal, A. (eds) *Indian Ocean Studies: Cultural, Social and Political Perspectives*, New York: Routledge.

Rangan, H. and Kull, C. 2009b. What Makes Ecology 'Political'?: Rethinking 'Scale' in Political Ecology. *Progress in Human Geography*, 33, 28–45.

Richardson, D. M. 2011. *Fifty Years of Invasion Ecology: The Legacy of Charles Elton*, Oxford, Blackwell Publishing.

Robbins, P. 2001. Tracking Invasive Land Covers in India, or Why Our Landscapes Have Never Been Modern. *Annals of the Association of American Geographers*, 91, 637–659.

Robbins, P. 2004a. Comparing Invasive Networks: Cultural and Political Biographies of Invasive Species. *The Geographical Review*, 94, 139–156.

Robbins, P. 2004b. Culture and Politics of Invasive Species. *The Geographical Review*, 94, iii–iv.

Robin, L. and Steffan, W. 2007. History for the Anthropocene. *History Compass*, 5, 1694–1719.

Rose, D. B., van Dooren, T., Chrulew, M., Cooke, S., Kearnes, M. and O'Gorman, E. 2012. Thinking through the Environment, Unsettling the Humanities. *Environmental Humanities*, 1, 1–5.

Rotherham, I., Rotherham, I. D., Lambert, R. A., and International Institute for Environment and Development (eds) 2011. *Invasive and Introduced Plants and Animals: Human Perceptions, Attitudes and Approaches to Management*, Washington, DC, Earthscan.

Simberloff, D. and Rejmánek, I. 2011. *Encyclopedia of Biological Invasions*, Chicago, University of Chicago Press.

Smout, T. C. 2003. The Alien Species in 20th-Century Britain: Constructing a New Vermin. *Landscape Research*, 28, 11–20.

Smout, T. C. 2011. How the Concept of Alien Species Emerged and Developed in 20th-Century Britain. In: Rotherham, I., Rotherham, I. D., Lambert, R. A., and International Institute for Environment and Development (eds) *Invasive and Introduced Plants and Animals: Human Perceptions, Attitudes and Approaches to Management*, Washington, DC: Earthscan.

Subramaniam, B. 2001. The Aliens Have Landed! Reflections on the Rhetoric of Biological Invasions. *Meridians*, 2, 26–40.

Tigger, D., Mulcock, J., Gaynor, A. and Toussaint, Y. 2008. Ecological Restoration, Cultural Preferences and the Negotiation of 'Nativeness' in Australia. *Geoforum*, 39, 1273–1283.

Whatmore, S. 2002. *Hybrid Geographies: Natures, Cultures, Spaces*, London, Sage.

Film

Dante, J., director, 1984. *Gremlins*, Warner Bros.

PART II

Invasion and the Anthropocene

2

BACK STORY

Migration, assimilation and invasion in the nineteenth century[1]

Harriet Ritvo

People were on the move in the nineteenth century. Millions of men and women took part in the massive transfers of human population that occurred during that period, spurred by war, famine, persecution, the search for a better life, or (most rarely) the spirit of adventure. The largest of these transfers—although by no means the only one—was from the so-called Old World to the so-called New. This is a story that has often been told, though its conclusion has been subject to repeated revision. That is to say, the consequences of these past population movements continue to unfold throughout the world, even as new movements are super-imposed on them. Of course, people are not unique in their mobility, as they are not unique in most of their attributes. Other animals share our basic desires with regard to prosperity and survival, and when they move independently, they are therefore likely to have similar motives. But, like people, they don't always move independently. And, as in the human case, when the migrations of animals are controlled by others, their journeys also reveal a great deal about those who are pulling the strings. A couple of animal stories can serve as examples. They both concern creatures transported far from their native habitats by the Anglophone expansions of the nineteenth century. The motives for their original introductions a century and a half ago were rather different, as have been their subsequent fates, but they were introduced to the same widely separated shores under circumstances that resembled each other in suggestive ways.

One story concerns the English or house sparrow (*Passer domesticus*), which was apparently first introduced into the United States by a nostalgic Englishman named Nicolas Pike in 1850, and subsequently reintroduced in various locations in eastern North America. In Darwinian terms, this was the beginning of a great success story. So conspicuously did the English sparrow flourish that in 1889, the Division of Economic Ornithology and Mammalogy (part of the U.S. Department of Agriculture—an ancestor of the current Fish and Wildlife Service) devoted its first monograph to it (Barrow, 1889; Moulton *et al.*, 2010). By 1928, a Department of Agriculture survey of introduced birds made the same point by opposite means, explaining the brevity of its entry on the species on the grounds that it 'receives such frequent comment that it requires no more than passing notice here' (Phillips, 1928: 49). It remains one of the commonest birds in North America, though its populations have recently suffered precipitous declines elsewhere in the world.

The sparrow's adaptation to North America may have been a triumph from the passerine point of view, but hominids soon came to a different conclusion. Although the first introduction was at mid-century, the most celebrated one occurred a decade and a half later. The *New York Times* chronicled the evolving opinions inspired by the new immigrants. In November 1868, it celebrated the 'wonderfully rapid increase in the number of sparrows which were imported from England a year or so ago'; they had done 'noble work' by eating the inchworms that infested the city's parks, described by the *Times* as 'the intolerable plague or numberless myriads of that most disgusting shiver-producing, cold-chills-down-your-back-generating, filthy and noisome of all crawling things'. The reporter praised the kindness of children who fed the sparrows and that of adults who subscribed to a fund that provided birdhouses for 'young married couples'; he promised that, if they continued to thrive and devour, English sparrows would be claimed as 'thoroughly naturalized citizens' (Anon, 1868: 8).

Two years later, sympathy was still strong, at least in some quarters. For example, the author of an anonymous letter to the editor of the *Times* criticized his fellow citizens in general, and Henry Bergh, the founder of the American Society for the Prevention of Cruelty to Animals, in particular, for failing to provide thirsty sparrows with water. Bergh took the allegation seriously enough to compose an immediate reply, pointing out that despite his 'profound interest . . . in all that relates to the sufferings of the brute creation—great and small,' neither he nor his society had authority to erect fountains in public parks (Anon, 1870a: 2; Bergh, 1870: 3). But the tide was already turning. Only a few months later the *Times* published an article entitled, 'Our Sparrows. What They Were Engaged to Do and How They Have Performed Their Work. How They Increase and Multiply—Do They Starve Our Native Song-Birds, and Must We Convert Them into Pot-Pies?' (Anon, 1870b: 6).

While the English sparrow was making itself at home in New York and adjoining territories, another creature was having a very different immigrant experience far to the southwest. In the early 1850s, after the American annexation of what became Texas, California, Arizona, and New Mexico, the U.S. Army found that patrolling the vast empty territory along the Mexican frontier was a daunting task, especially in the overwhelming absence of roads. The horses and mules that normally hauled soldiers and their gear did not function efficiently in this harsh new environment. Of course, though the challenges of the desert environment were new to the U.S. Army, they were not absolutely new. The soldiers and merchants of North Africa and the Middle East had solved a similar problem centuries earlier, and some open-minded Americans were aware of this (see Bulliet, 1990).[2] Several officials serving in the dry trackless regions therefore persuaded Jefferson Davis, then the U.S. Secretary of War, that what the army needed was camels (Figure 2.1), and in 1855 Congress appropriated $30,000 to test the idea (Marsh, 1856: 210).

Acquiring camels was more expensive than acquiring sparrows, partly because they are much larger and partly because such transactions required intermediate negotiations with people, including camel owners, foreign government, and customs officials. And the animals themselves demanded significantly more attention, which Americans familiar only with such northern ungulates as horses and cattle were ill equipped to provide. In consequence, a Syrian handler named Hadji Ali (soon anglicized to 'Hi Jolly') was hired to accompany the first shipment of camels; he outlasted his charges and was ultimately buried in Quartzsite, Arizona, where his tomb, which also commemorates the original Camel Corps, now constitutes the town's primary tourist attraction.[3] A total of 75 camels survived their

FIGURE 2.1 Camels.

Source: Goodrich (1861: 576).

ocean voyages and their subsequent treks to army posts throughout the southwest. The officers who used them on missions were, on the whole, favorably impressed, while the muleteers who took care of them tended to hold them in more measured esteem.

But these discordant evaluations did not explain the ultimate failure of the experiment. With the outbreak of the Civil War, responsibility for the camels, whose numbers had grown somewhat through natural increase, passed to the Confederacy. Even their early advocate Jefferson Davis had other priorities at that point. Some of the camels were sold to circuses, menageries, and zoos; others were simply allowed to wander away into the wild dry lands. They were sighted (and chased and hunted) with decreasing frequency during the post-war decades (Perrine, 1925). In 1901, a journalist who considered the whole episode to be 'one of the comedies that may once in a while be found in even the dullest and most ponderous volumes of public records from the Government Printing Office' reported that 'now and then a passenger on the Southern Pacific Railroad . . . has had a sight of some gaunt, bony and decrepit old camel . . . grown white with age, [and] become as wild and intractable as any mustang' (Griswold, 1901: 218–219).

Of course, the details of the assimilation or attempted assimilation—how many individuals were involved, whether they were wild or domesticated, where they went and where they came from, whether the enterprise succeeded or failed— made a great difference to the imported creatures as well as to the importers. Such attempts, often termed 'acclimatization', became relatively frequent during the nineteenth century, though the simple desire to acclimatize was the reverse of novel. Whether so labeled or not, acclimatization has been a frequent corollary of domestication, as useful plants and animals have followed human routes of trade and migration; it thus dates from the earliest development of agriculture, 10,000 years and more ago. Indeed, much of the history of the world, at least from the perspective of environmental history, can be understood in terms of the dispersal and acclimatization of livestock and crops.

Historically and prehistorically, people have taken animals and plants along with them in order to re-establish their pastoral or agricultural way of life in a new setting. Thus the bones of domesticated animals (and the seeds and other remains of domesticated plants) can help archaeologists trace, for example, the spread of Neolithic agriculture from the various centers where it originated. (The agricultural complex that was ultimately transferred throughout the temperate world by European colonizers in the post-Columbian period, based on cattle, sheep, and goats, along with wheat, barley, peas, and lentils, was derived ultimately from the ancient farmers of the eastern Mediterranean.) Even the remains of less apparently useful (or at any rate, less edible) domesticated animals can signal human migration patterns. For example, the prevalence of orange cats in parts of northwestern Europe indicates long ago Viking settlement, and the relative frequency (greater than further south and decreasing toward the Pacific) of robust polydactyl cats (a mutation that apparently arose in colonial Boston) along the northern range of American states indicates the westward movement of New Englanders (Todd, 1977: 100–107).

Alfred Crosby has christened the process by which this assemblage of domesticated animals and plants (along with the weeds, pests, and diseases that inevitably
accompanied them) achieved their current global range, 'ecological imperialism',
replacing or subsuming his earlier coinage, 'the Columbian exchange' (see Crosby,
1986, 1972). These labels are somewhat inconsistent in their political implications,
but they both have validity. Especially with regard to plants, the Americas have
transformed the rest of the world at least as much as they have been transformed
by it: corn (maize) and potatoes are now everywhere. But of course American
imperialism, when it emerged, did not result from this multidirectional dissemina-
tion of indigenous vegetables. Instead it was a consequence of the final westward
transfer of the combination of domesticated plants and animals initially developed
in ancient southwest Asia, and gradually adapted to the colder wetter climates of
northern Europe and eastern North America.

The instigators of the wave of acclimatization attempts that crested in the late
nineteenth century often claimed that their motives were similarly utilitarian. But
as is often the case, their actions told a somewhat different story. The American
experiences of the English sparrow and the camel suggest the much smaller scale
of such transfers, though the relatively few imported sparrows ultimately populated an entire continent through their own vigorous efforts. In addition, most
nineteenth-century introductions resulted from the vision or desire of a few
individuals, not an entire community or society; they involved the introduction of
more or less exotic animals to that community, rather than the transportation by
human migrants of familiar animals along with tools and household goods in order
to reestablish their economic routine. Self-conscious efforts at acclimatization also
embodied assumptions and aspirations that were much more grandiose and self-
confident: the notion that nature was vulnerable to human control and the desire
to exercise that control by improving extant biota. In many ways acclimatization
efforts seemed more like a continuation of a rather different activity, which also
had ancient roots, though not quite as ancient: the keeping of exotic animals in
game parks and private menageries (for the rich), and in public menageries and
sideshows (for the poor). This practice similarly both reflected the wealth of human
proprietors, and implicitly suggested a still greater source of power, the ability to
categorize and re-categorize, since caged or confined creatures—even large dangerous ones like tigers or elephants or rhinoceroses—inevitably undermine the
distinction between the domesticated and the wild.

The scale of these nineteenth-century enterprises was often paradoxical: they
simultaneously displayed both hubristic grandeur in their aspirations and narrow
focus and limited impact in their realizations. For example, the thirteenth Earl of
Derby, whose estate at Knowsley, near Liverpool, housed the largest private
collection of exotic wild animals in Britain, was one of the founders of the
Zoological Society of London and served as its President from 1831 until he died
in 1851. He bankrolled collecting expeditions to the remote corners of the world,
and there were frequent exchanges of animals between his Knowsley menagerie
and the Zoo at Regent's Park, as well as other public collections (Fisher and

Jackson, 2002: 44–51). These exchanges were by no means unequal; indeed the Earl's personal zoo was decidedly superior. At his death it covered more than 100 acres and included 318 species of birds (1272 individuals) and 94 species of mammals (345 individuals) (Fisher, 2002: 85–86). Among its denizens were bison, kangaroos, zebras, lemurs, numbats, and llamas, as well as many species of deer, antelope, and sheep. In addition to providing his animals with food, lodging, and expert veterinary attention (sometimes from the most distinguished human specialists), Derby had them immortalized by celebrated artists (including Edward Lear) when they were alive, and by expert taxidermists afterwards. But he made no plans for his menagerie, or even for any of the breeding groups it contained, to survive him. His heir, already an important politician and soon to be prime minister, had no interest in the animals and sold them at auction as soon as possible.

Late in the century, the eleventh Duke of Bedford, also a long-serving President of the Zoological Society of London (1899–1936), established a menagerie at Woburn Abbey, his Bedfordshire estate. By this time, the rationale for accumulating such a vast private collection of living animals had evolved. The Woburn park contained only ungulates (and a few other grazers, like kangaroos and wallabies): its residents included various deer, goats, cattle, gazelles, antelope, tapirs, giraffes, sheep, zebras, llamas, and asses. A summary census printed in 1905 made it clear that, unlike his distinguished predecessor, the Duke collected with a view to acclimatization. 'Only those animals believed to be hardy' were selected for trial, and animals that were not 'good specimens', either because of their savage dispositions or because their constitutions were not well adapted to the environment of an English park, did not survive long (Anon, 1905).

That is to say, he collected with a view to the future, hoping that his park would serve as a way station for species that might find new homes in Britain, whether in stockyards or on public or private display. In several cases, Woburn Abbey in fact provided a refuge—or even the last refuge—for remnant populations. Before the Boxer Rebellion, the Duke secured a small herd of Père David's deer, a species otherwise exclusively maintained in the imperial parks of China (and so already extinct in the wild). An original herd of 18 had grown to 67 by 1913 (Chalmers Mitchell, 1913: 79). Since their Chinese relatives fell victim to political turmoil, all the current members of the species descend from the Woburn herd. He also nurtured the Przewalski's horse—a rare wild relative of domesticated horses and ponies, discovered (at least by European science) only in the late nineteenth century, when it was on the verge of extinction (R. L., 1901: 103).

The Duke's emphasis on preservation also echoed a shift that was to become increasingly evident in the rhetoric of zoological gardens in the course of the twentieth century. As zoo-goers will have noticed, preservation, both of individual animals and of threatened species, has loomed increasingly large in their publicity, though, of course, intention is often one thing, and results are another. Less predictive of the evolution of zoo policies was the Duke's emphasis on acclimatization. His menageries contained mostly ungulates because those are the animals that people like to eat. Although there have been occasional deviations, such as the scandal that engulfed the Atlanta Zoo in 1984, when it emerged that 'a city worker was making rabbit stew and other dishes out of the surplus small animals he had bought from the zoo's children's exhibit',[4] on the whole, modern zoos have taken care not to suggest that their charges, or the offspring of their charges, will end their days on someone's plate.

But this distinction—between natural history and agriculture, to put it one way—seemed less important in the early days of public zoos. Indeed, it hardly existed. On the contrary, the first goal mentioned in the 'Prospectus' of the Zoological Society of London was to introduce new varieties of animals for 'domestication or for stocking our farm-yards, woods, pleasure grounds and wastes' (Bastin, 1970: 385). To this end, along with the menagerie at Regent's Park, the young society established a breeding farm at Kingston Hill, not far to the west of London. It lasted only a few years, as the market for the stud services of zebus and zebras turned out to be small. But the notion that the zoo could supplement or enhance the British diet persisted, at least in some particularly active imaginations. Frank Buckland, an eccentric and omnivorous naturalist, successfully requested permission to cook and eat the remains of the zoo's deceased residents. Among the species he (and his unfortunate dinner guests) sampled were elephant, giraffe, and panther (that is, leopard) (Ritvo, 1987: 237–241).

Naturalists like Buckland, along with wealthy owners of private menageries, founded the Society for the Acclimatisation of Animals, Birds, Fishes, Insects and Vegetables within the United Kingdom in 1860. They were following in the footsteps of French colleagues, who had founded the Société Zoologique d'Acclimatation in 1854. But their proximate inspiration was a zoological dinner held at a London tavern in 1859, at which the gathered naturalists and menagerists enjoyed the haunch of an eland descended from the Earl of Derby's herd at Knowsley Park. The declared objects of the society were grandiose and diffuse: to introduce, acclimatize, and domesticate 'all innocuous animals, birds, fishes, insects, and vegetables, whether useful or ornamental'; to perfect, propagate and hybridize these introductions; to spread 'indigenous animals, &c.' within the United Kingdom; to procure 'animals &c., from British Colonies and foreign countries'; and to transmit 'animals, &c. from England to her colonies and foreign parts'. If all these objects had been achieved, the result would have been a completely homogenized globe, at least with respect to the flora and the fauna. In fact, of course, none of them came close to realization. Despite Buckland's ambitious wish list, which included beavers and kangaroos, along with the more predictable bovids and cervids, most society members confined their attention to a scattering of birds and sheep, none of which made much impact on the resident plants and animals, whether wild or domesticated. The Society itself survived only through 1866, when it enrolled only 270 members, of which 90 were life members who had therefore lost the power of expressing disaffection; it was then absorbed by the Ornithological Society of London (Lever, 1977: 29–35; Ritvo, 1987: 239; see also Lever, 1992).

The French society was larger (2600 members in 1860, including a scattering of foreign dignitaries), longer-lasting, and more firmly grounded, both in Paris, where it controlled its own Jardin d'Acclimatation, and within a network of colonial societies (Anderson, 1992: 143–144; Osborne, 2000: 143–145). It kept elaborate records, which could be consulted by any landowner wishing to diversify his livestock. But, like those in Britain, French acclimatization efforts never had a significant local economic effect, nor did they transform the landscape. Instead they made life a little more curious and entertaining. By the end of the century, the generalization that 'animal acclimatisation in Europe is now mainly sentimental or is carried out in the interests of sport or the picturesque' applied in France as well as Britain, where, according to a commentator in the *Quarterly Review*, aficionados of the exotic could savor 'the pleasure of watching [the] unfamiliar forms [of Japanese apes and American prairie dogs, as well as gazelles and zebras] amid the familiar scenery' (Anon, 1900: 199–201).

The main economic impact of French acclimatization efforts was in such warmer colonial locations as Algeria. And though the British society lacked official or quasi-official support (at least with regard to animals—Kew Gardens was at the center of a network concerned with the empire-wide distribution of plants that might produce economic benefits), the Anglophone acclimatization movement also had great (though not necessarily similar) impact outside the home islands. Acclimatization societies quickly sprang up throughout Australia and New Zealand, where members embraced a weightier mission than the one undertaken by Frank Buckland or the Duke of Bedford. They felt that new kinds of animals were not needed merely for aesthetic or culinary diversification; they were needed to repair the defects of the indigenous faunas, which lacked the 'serviceable animals' found so abundantly in England, including, among others, the deer, the partridge, the rook, the hare, and the sparrow. The heavy medals struck in 1868 by the Acclimatisation Society of Victoria give a sense of the seriousness with which they approached this endeavor. One side featured a wreath of imported plants, surrounding the society's name, the other a group portrait of a hare, a swan, a goat, and an alpaca, among other desirable exotic animals.[5]

Their passion was rooted in a perception of dearth. Acclimatizers complained that while nature had provided other temperate lands with 'a great profusion . . . of ruminants good for food, *not one single creature of the kind inhabits Australia!*' They were not discouraged when immigrant rabbits and sparrows began to despoil gardens and fields, merely suggesting the hair of the dog as remedy: it might be advisable to 'introduce the mongoose to war against the rabbits'. They continued to urge 'the acclimatization of every good thing the world contains' until 'the country teemed with animals introduced from other countries'.[6]

As was often the case, ordinary domesticated animals were not of primary concern to the most enthusiastic and visionary acclimatizers, though in many places cattle and sheep were more influential than rabbits or rats or sparrows in converting alien landscapes into homelike ones. But, in Australia, as in Texas and Arizona, extraordinary domesticated animals could fall into another category. Similar problems—vast trackless deserts that nevertheless required to be traversed by people and their equipment—suggested similar solutions. A few immigrant camels arrived in Australia in 1840, but the ship of the desert was not integrated into the economic life of the colony (or colonies) for several decades (see Rangan and Kull, 2009). In the 1860s, just as the Civil War deflected official interest from the American camels, their Australian conspecifics were beginning to flourish, their manifest utility outweighing the perception of some who used them, that they could be spiteful, sulky, and insubordinate (Winnecke, 1884: 1–5). They even received appreciative notice in the imperial metropolis: by 1878, *Nature* reported approvingly that they worked well when yoked in pairs like oxen, and that they remained very useful in exploring expeditions, though most labored in the service of ordinary commercial purposes (Anon, 1878: 337). They also carried materials for major infrastructure projects that brought piped water and the telegraph to the dry interior. A camel breeding stud was established in 1866; overall, in addition to homegrown animals, approximately 10,000 to 12,000 camels were imported for draft and for riding during the subsequent half century.[7] Their importance continued until the 1920s, when they were supplanted by cars and trucks—the same fate that had already befallen horses in Europe and elsewhere.

Suddenly, what had seemed an unusually successful adventure in acclimatization took on a different cast. As in the American southwest, once the camels lost their utility, they became completely superfluous. A camel-sized pet is an expensive luxury, and there was no significant circus or zoo market for animals that had long ceased to be exotic. So some were shot and others were set free to roam by kinder-hearted owners. At this point the Australian story diverged from the American one once again. Camels had lived in Australia for at least as long as many of its human inhabitants (that is, the ones with European roots) in terms of years, and in terms of generations, they had lived there longer. They were well adapted to the harsh terrain, where they foraged and reproduced, rather than dwindling and dying. As of 2009, according to the Australian Government, their feral descendants numbered close to one million—by far the largest herd of free-living camels in the world; a year later the *Meat Trade News Daily* estimated the camel population at 1.2 million.[8] They competed for resources with other animals, wild and domesticated, and it was feared that they were disrupting fragile desert ecosystems. Like some of the elephant populations of south and southeast Asia, they were occasionally reported to terrorize small towns. After helping to build the nation, they had, it was asserted, 'outstayed their welcome'.[9] At least until recently, culling did not keep up with new births; and the market for camel meat that had arisen in the 1980s made even less of a dent. Unsurprisingly, in a pattern that had also emerged with regard to feral horses, burros, and pigs in North America, as officials contemplated more drastic methods that would quickly reduce the population by two-thirds, human resistance also emerged, whether based on regard for the welfare of individual camels, the hope the camels could be converted dead or alive into a profit center (meat or tourism), or the fear that large-scale eradication would require the violation of property rights.[10]

The acclimatization agenda in New Zealand was somewhat different with regard to its objects, but at least equally enthusiastic and even more persistent. Since the topography and climate of New Zealand differ greatly from those of Australia, and so camels were never at the top of the list of targets for introduction. But acclimatizers in both places shared the desire to convert their new homelands into the most plausible possible simulacra of their old ones. In the initial burst of enthusiasm, as elsewhere, animal introductions were scattershot—anything that appealed to individual acclimatizers. But soon the focus shifted to the re-creation half a world away of the staples of British outdoor sport: deer, game birds like pheasants and grouse, and game fish like trout and salmon. Some of these thrived, with a transformative effect on the local fauna, and others languished. The ubiquitous local societies attempted to protect them by eliminating indigenous predators. In 1906, for example, the Wellington Acclimatisation Society was taking measures to combat 'the shag menace to trout'.[11] In the course of the twentieth century new perspectives on this practice emerged and enthusiasm for acclimatization diminished—though not everywhere. The plaque on an imposing monument to trout acclimatization reads:

> This centennial plaque was presented to the Auckland Acclimatisation Society to convey the gratitude of past, present, and future generations of trout anglers in New Zealand for the society's successful importation of Californian rainbow trout ova in 1883, its hatching of the eggs in the Auckland Domain Pond and its subsequent distribution of the fish and their progeny to many New Zealand waters.[12]

In 1990, the local societies were abolished; that is to say, they were converted into fish and game councils.[13]

These examples demonstrate that utility, like many other things, is a matter of perspective. Because frivolous (or worse) as they may seem from a contemporary vantage point, the instigators of all these acclimatization attempts understood themselves to be acting in the public interest, and not just for their own idiosyncratic satisfaction.

Perhaps the most poignant demonstration of this is another well-known American saga, that of the introduction of the starling (Figure 2.2). The starting point was also New York City, the scene of the excessively successful sparrow release. In 1871, the American Acclimatization Society was founded to provide a formal institutional base for such attempts. It is widely reported, though occasionally doubted, that its moving spirit, a prosperous pharmacist named Eugene Schieffelin, wished to introduce to the United States all the birds named in Shakespeare. One reason for doubt is simply quantitative—according to a little book called *The Birds of Shakespeare*, which was published in 1916, that tally would include well over 50 species, not all of them native to Britain (Geikie, 1916). But nevertheless this notion is persistent—thus a recent article on this topic in *Scientific American* was headlined 'Shakespeare to Blame for Introduction of European Starlings to U.S.'(Mirsky, 2008). Less controversially, this attempt—which also turned out to be excessively successful—was part of what the Department of

COMMON AND BLACK STARLING (⅓ nat. size).

FIGURE 2.2 Starlings.

Source: Lydekker (1894–1895: Vol. III: 345).

Agriculture retrospectively characterized as 'the many attempts to add to our bird fauna the attractive and familiar [and "useful"] song birds of Europe' (Phillips, 1928: 48–49). The report of the 1877 annual meeting of the American Acclimatization Society, at which the starling release was triumphantly announced, also approvingly noted more or less successful releases of English skylarks, pheasants, chaffinches, and blackbirds, and looked forward to the introduction of English titmice and robins, as well as additional chaffinches, blackbirds, and skylarks—all characterized as 'birds which were useful to the farmer and contributed to the beauty of the groves and fields' (Anon, 1877: 2).

The acclimatization project has often been interpreted as a somewhat naïve and crude expression of the motives that underlay nineteenth-century imperialism—intellectual and scientific, as well as political and military—more generally. This understanding is compelling, but not necessarily comprehensive. There is, for one thing, a significant difference between the imposition of the European biota on the rest of the world, and the transfer of exotic animals and plants to the homeland (whether inherited or adopted). And for another, the enterprise of acclimatization is much more likely to demonstrate the limitations of human control of nature than the reverse—whether the targets of acclimatization shrivel and die, or whether they reproduce with unanticipated enthusiasm. Already in the nineteenth century, introduction of exotic plants and animals could be seen as a kind of Pandora's box, at least when they were imported into Europe or heavily Europeanized colonies or ex-colonies. For example, to return to eastern North America, the Society for the Protection of Native Plants (now the New England Wild Flower Society) was founded in 1900, in order to 'conserve and promote the region's native plants'.[14] It was the first such organization in the United States, but in the intervening century societies with similar goals have been established across the continent. The commitment to preserve native flora and fauna from the encroachment of aliens marked a turn, conscious or otherwise, from offense to defense—perhaps in the American context, to be read in conjunction with the Chinese Exclusion Act of 1882 or the more comprehensive Immigration Act of 1924. And of course the American context was not the only relevant one, in the nineteenth century or later; elsewhere the defense of the native would become still more strenuous.

Notes

1 A shorter version of this chapter was originally published as 'Going Forth and Multiplying: Animal Acclimatization and Invasion', in *Environmental History* 17 (2012), 404–414.
2 Bulliet (1990) gives a definitive account of the integration of camel transport into the economies and societies of the Middle East and North Africa.
3 See Woodbury, 'U.S. Camel Corps remembered in Quartzsite, Arizona' *Out West* (2003).
4 See Schmidt, 'Civic Leaders Planning Reforms for Atlanta Zoo' (1984), *New York Times* (August 28 1984).
5 For an image of the medal, see: http://museumvictoria.com.au/collections/items/76618/medal-acclimatisation-society-of-victoria-bronze-australia-1868.
6 Acclimatisation Society of Victoria, *First Annual Report* (1862), 8, 39 and *Sixth Annual Report* (1868), pp. 29–30; South Australian Zoological and Acclimatization Society, *Seventh Annual Report* (1885), 7; Acclimatisation Society of Victoria, *Third Annual Report* (1864), p. 30 and *Fifth Annual Report* (1867), p. 25.
7 'Camels Australia Export', available at: http://www.camelsaust.com.au/history.htm (accessed March 30 2011); 'A Brief History of Camels in Australia', based on *Strategies for Development* (1993) prepared by the Camel Industry Steering Committee for the Northern Territory Government, available at: http://camelfarm.com/camels/camels_australia.html.
8 'Australia: the world's largest camel population', *Meat Trade News Daily*, 5 September 2010. Available at: http://www.meattradenewsdaily.co.uk/news/100910/australia___the_worlds_largest_camel_population_.aspx (accessed May 18 2012); 'Camel fact sheet', Department of the Environment, Water, Heritage and the Arts, 2009. Available at: http://www.environment.gov.au/biodiversity/invasive/publications/camel-factsheet.html (accessed May 18, 2012).
9 'A million camels plague Australia', *National Geographic News*, October 26 2009. Available at: http://news.nationalgeographic.com/news/2009/10/091026-australia-camels-video-ap.html (accessed March 30 2011).

10 'Feral camels in Western Australia', Department of Environment and Conservation, Western Australia, 2009. Available at: http://www.dec.wa.gov.au/content/view/3224/1968/ (accessed March 30 2011).
11 *The Press*, Christchurch, for 23 February 1906. Reproduced in 'Acclimatisation Societies in New Zealand'. Available at:http://www.pyenet.co.nz/familytrees/acclimatisation/#pyeaccshagmenace (accessed April 1 2011).
12 See http://www.tongarirorivermotel.co.nz/trout-fishing/history-of-taupo-fishery/.
13 Conservation Law Reform Act 1990 031 Legislation NZ. Available at: http://legislation.knowledge-basket.co.nz/gpacts/public/text/1990/se/031se74.html (accessed April 1 2011).
14 New England Wild Flower Society website. Available at: http://www.newfs.org/about/history/?searchterm=history (accessed April 1 2011).

References

Anderson, W. 1992. Climates of Opinion, Acclimatization in 19th-Century France and England + Victorian Botanical and Biological Thought. *Victorian Studies*, 35, 134–157.
Anon, 1868. Our Feathered Friends. *New York Times*, November 22.
Anon, 1870a. Man's Inhumanity to Birds. *New York Times*, July 22.
Anon, 1870b. Our Sparrows. *New York Times*, November 20.
Anon, 1877. American Acclimatization Society. *New York Times*, November 15.
Anon, 1878. Geographical Notes. *Nature*, 337.
Anon, 1900. New Creatures for Old Countries. *Quarterly Review*, 192, 199–200.
Anon, 1905. A Record of the Collection of Foreign Animals Kept by the Duke of Bedford in Woburn Park 1892 to 1905.
Barrow, W. B. 1889. *The English Sparrow (Passer domesticus) in North America in its Relations with Agriculture*, Washington, DC: United States Department of Agriculture Division of Economic Ornithology and Mammalogy.
Bastin, J. 1970. The First Prospectus of the Zoological Society of London: A New Light on the Society's Origins. *Journal of the Society for the Bibliography of Natural History*, 5, 385.
Bergh, H. 1870. Mr Bergh and the Sparrows; A Defence Against Certain Aspersions. *New York Times*, July 23.
Bulliet, R. W. 1990. *The Camel and the Wheel*, New York, Columbia University Press.
Chalmers Mitchell, P. 1913. Zoological Gardens and the Preservation of Fauna. *Nature*, 90, 79.
Crosby, A. W. 1972. *The Columbian Exchange: Biological and Cultural Consequences of 1492*, Westport, CT, Greenwood Publishing Co.
Crosby, A. W. 1986. *Ecological Imperialism: The Biological Expansion of Europe*, Cambridge, Cambridge University Press.
Fisher, C. 2002. The Knowsley Aviary & Menagerie. In: Fisher, C. T. (ed.) *A Passion for Natural History: The Life and Legacy of the 13th Earl of Derby*, Liverpool: National Museums and Galleries of Merseyside.
Fisher, C. and Jackson, C. E. 2002. The Earl of Derby as Scientist. In: Fisher, C. T. (ed.) *A Passion for Natural History: The Life and Legacy of the 13th Earl of Derby*, Liverpool: National Museums and Galleries of Merseyside.
Geikie, A. 1916. *The Birds of Shakespeare*, Glasgow, J. Maclehouse.
Goodrich, S. G. 1861. *Illustrated History of the Animal Kingdom, Being a Systematic and Popular Description of the Habits, Structure and Classification of Animals from the Highest to the Lowest Forms, with their Relations to Agriculture, Commerce, Manufactures, and the Arts*, New York, Derby and Jackson.
Griswold, H. T. 1901. The Camel Comedy. *Current Literature*, 31, 218–219.
Lever, C. 1977. *The Naturalized Animals of the British Isles*, London, Hutchinson.

Lever, C. 1992. *They Dined on Eland: The Story of the Acclimatisation Societies*, London, Quiller Press.

Lydekker, R. (ed.) 1894–1895. *The Royal Natural History*, Vol. III. London, Frederick Warne.

Marsh, G. P. 1856. *The Camel: His Organization Habits and Uses Considered with Reference to his Introduction into the United States*, Boston, Gould and Lincoln.

Mirsky, S. 2008. Shakespeare to Blame for Introduction of European Starlings to U.S. *Scientific American*, May 23.

Moulton, M. P., Cropper Jr, W. P., Avery, M. L. and Moulton, L. E. 2010. The Earliest House Sparrow Introductions to North America. *Biological Invasions*, 12, 2955–2958.

Osborne, M. A. 2000. Acclimatizing the World: A History of the Paradigmatic Colonial Science. *Osiris*, 15, 135–151.

Perrine, F. S. 1925. Uncle Sam's Camel Corps. *New Mexico Historical Review*, 1, 434–444.

Phillips, J. C. 1928. *Wild Birds Introduced or Transplanted in North America*, Washington, DC: United States Department of Agriculture.

Rangan, H. and Kull, C. 2009. The Indian Ocean and the Making of Outback Australia: An Ecocultural Odyssey. In: Moorthy, S. and Jamal, A. (eds) *Indian Ocean Studies: Cultural, Social and Political Perspectives*, New York: Routledge.

Ritvo, H. 1987. *The Animal Estate: The English and Other Creatures in the Victorian Age*, Cambridge, MA, Harvard University Press.

Ritvo, H. 2012. Going Forth and Multiplying: Animal Acclimatization and Invasion. *Environmental History*, 17, 404–414.

R. L. 1901. Przewalski's Horse at Woburn Abbey. *Nature*, 65, 103.

Schmidt, W. E. 1984. Civic Leaders Planning Reforms for Atlanta Zoo. *New York Times* (August 28 1984). Available at: http://www.nytimes.com/1984/08/28/us/civic-leaders-planning-reforms-for-atlanta-zoo.html?scp=1&sq=atlanta%20zoo%20%20rabbit%20sale&st=cse.

Todd, N. B. 1977. Cats and Commerce. *Scientific American*, 237, 100–107.

Winnecke, C. 1884. *Mr Winnecke's Explorations During 1883*, Adelaide, Govt Printer.

Woodbury, C. 2003. U.S. Camel Corps Remembered in Quartzsite, Arizona. *Out West*. Available at: http://www.outwestnewspaper.com/camels.html (accessed March 27 2011).

3

FRAGMENTS FOR A POSTCOLONIAL CRITIQUE OF THE ANTHROPOCENE

Invasion biology and environmental security

Gilbert Caluya

Despite the tenuous status of the concept of the 'Anthropocene' within the field of geology, the concept has nevertheless garnered growing interest from the humanities and social sciences. The Anthropocene is used to mark a period of time when human activity has drastically changed the face of the world such that it approaches a geological force. As a concept, it helps us to grasp the immensity of the problem of climate change, to grapple with its global nature, and it acts as a banner to rally us into responsibility. Some argue that this instantiates a new humanities or, at the very least, that the humanities must now undo much of their prior thinking to meet this call for responsibility. Yet the humanities have long argued for a more environmentally conscious approach to life and living. So what precisely do the humanities and social sciences gain from the concept of the 'Anthropocene'? Or, more importantly, what gets left out of the picture when we use this lens?

This chapter argues that how we go about managing and protecting nature can often reinscribe human systems of oppression and discrimination, drawing on invasion biology as its example. The first section introduces the concept of the Anthropocene while tracing its movement from the sciences to the humanities, where it has been recently posed by Dipesh Chakrabarty (2012) as a 'challenge' to postcolonial studies. While in general I agree with Chakrabarty's claim that 'the human' under the Anthropocene theoretically challenges the human conceptualised in the Enlightenment and in postcolonialism respectively, I suggest that the former is liable to be co-opted by the Enlightenment version of the human if postcolonial critiques are not taken seriously. This argument is posed through the example of invasion biology to which the following sections turn. It begins by contextualising invasion biology within imperial ecology in order to trace the conceptual connections between nature and the human under colonial/imperial histories of oppression. The final section returns to the present, drawing on disparate cultural texts in Australia to show the continued imbrication of human and non-human biota management under the paranoid imaginary of border security logics, which rely upon the protectionist rhetoric of invasion.

The Anthropos in the Anthropocene

In 2000, Paul Crutzen, Nobel Prize-winning atmospheric chemist, used the term 'Anthropocene' to mark a distinction from the Holocene on the grounds that humans have radically changed the face of the world. Since then journal articles debating the 'Anthropocene' in the natural sciences have increased exponentially. In 2008, a proposal was presented to the Stratigraphy Commission of the Geological Society of London to evaluate the scientific validity of designating the 'Anthropocene' as a formal unit of geological time. The proposal was supported by a large majority at the conference and an Anthropocene Working Group of the Subcommission on Quarternary Stratigraphy was formed to assess the scientific evidence in favour of designating an Anthropocene Era. Several grant applications have been submitted on the subject, new research groups have sprung up around the world and a new collaboration between scientists and cultural/arts workers is currently planning two exhibitions in Germany (see Anthropocene Working Group, 2009).

Despite the buzz of activity[1] surrounding this new concept, the Anthropocene has yet to be formally accepted by the Geological Society of London (i.e. the scientific body responsible for determining geological timelines). As is evident from an e-mail exchange between Dr Jan Zalasiewicz (Chair of the Anthropocene Working Group) and Dr Davor Vidas, dated 28 August 2009, there is 'a reasonable general case made for considering the Anthropocene as a formal unit' but the 'term is not yet formal' and 'over the next few years [the Anthropocene Working Group] will examine and weigh the geological evidence' (Anthropocene Working Group, 2009: 7). According to their website, the group aims to have a proposal formalising the term 'Anthropocene' by 2016.[2]

Nevertheless, the uncertain status of the 'Anthropocene' in the scientific community has not dulled humanities and social sciences' enthusiasm for the term as evidenced by the growing number of associated conferences. In Australia, the Humanities Research Centre of Australian National University hosted the 2012 Annual Meeting of the Consortium of Humanities Centers and Institutes on 13–16 June 2012, which was dedicated to 'Anthropocene Humanities'. In 2013, a conference on 'Society in the Anthropocene' was held at Bristol University, another conference on 'The History and Politics of the Anthropocene' was held at Chicago University, and in Australia, the Animal Studies Group at the University of Sydney hosted a conference on 'Life in the Anthropocene'. In the same year, the second conference of the Environmental Humanities Network was themed 'Culture and the Anthropocene' and held on 14–16 June 2013 at the Rachel Carson Centre for Environment and Society in Munich.

Of course the Anthropocene Humanities (to borrow one of the conference titles) is simply the latest shift in the longer history of a broader environmental humanities, which I understand to include earlier ecopoetics and ecocriticism in literature (which have their roots in Romanticism), Marxist environmentalism, radical vegetarianism in feminism (Adams, 1990), post-scarcity anarchism in politics (Bookchin, 1971), and environmental security in international relations.[3] In recent years, this traditional interest in the environment within the humanities and social sciences has grown again: cultural anthropologists are turning to multispecies ethnographies, urban scholars are mapping the environmental destruction caused by cities and charting new pathways for eco-cities, while political economists have extend the notion of 'hegemony' beyond an historical bloc of a politico-economic system to incorporate a socio-ecological system (for multispecies ethnographies, see the collection of articles presented in the 2010 special edition of *Cultural Anthropology*, volume 25, issue 4; Hodson and Marvin, 2010; Phelan *et al.*, 2012).

In part, this new environmental humanities is the result of funding criteria that act as governmental incentives/imperatives to direct national research trajectories. For example, the UK's cross-council research theme, 'Living with Environmental Change', was the result of the Stern Review on the Economics of Climate Change, the United Nations' Millennium Ecosystem Assessment, and the reports by the Intergovernmental Panel on Climate Change, while Canada's Social Sciences and Humanities Research Council's Funding Priority 'Canadian Environmental Issues' was identified by the National Roundtable on the Environment and the Economy, and the Australian Research Council has as one of its National Research Priorities 'An Environmentally Sustainable Australia'. Funding imperatives aside, this does not detract from these theoretical developments as real responses to the emerging threat of climate change/global warming.

Yet it is also responding to perceived deficiencies in earlier theorising. Intellectually, the Anthropocene Humanities is supported by a growing movement away from the perceived excesses of the so-called 'linguistic turn' and 'post-structuralism' (always already inadequate Anglophonic catch-all terms for a multiplicity of French theories deploying rhetorical, etymological, grammatical, linguistic, semiotic and discursive approaches) towards the re-discovery of matter, things and the non-human featured in various theories: post-humanist, new materialist, Deleuzian and Actor-Network Theory. Within this wider context, new concepts have entered the humanities lexicon (e.g. 'the Anthropocene' and 'more-than-human'), while older concepts (such as Feuerbach's notion of 'species-being') are recycled and returned to the mantelpiece of fashionable theory.

At one level, what the Anthropocene does for humanities is give us a name for our period, a linguistic concretion of all that is environmentally wrong with our world, by bringing dispersed interests and politics together under climate change/global warming. As a concept, it helps us to grasp the immensity of the problem of climate change – to grapple with its global nature – and it acts as a banner to rally us into responsibility.

My concern with this turn is that poststructuralism[4] (often confused with anti-foundationalism and postmodernism in these debates) is increasingly becoming a new anachronism, whose increasingly naff formulations about 'discursive constitution' and 'historico-cultural contexts' are met with eyeball rolling at *de rigueur* conferences. Simultaneously, issues of identity have seemingly receded into the past, with 'identity politics' in particular increasingly portrayed as a period of navel-gazing victimisation. While these tendencies are not evident among the prominent authors of the field, the general rank-and-file have taken to this 'turn' as a substitute for previous approaches, theories, topics and methods. As one conference participant bluntly put it to me: 'Yes, yes, we all know about racism, but what about climate change?'

I hope the reader can forgive me my suspicions, but as a cultural scholar of race and racism I cannot help but feel that this has been a most opportune shift for some. For a long stretch of history, most people on this world had been relegated to the status of sub- or infra-humans, the fact of which was used to exclude or excise them from the realm not simply of political rights but also of full participation in social and economic life.[5] The political, and often violent, struggles (whether feminist, anti-racist or queer) to be incorporated in the historically narrow category of 'humanity' are quite recent in our collective history but have had limited and uneven success. Consequently, I am suspicious when just as the category of the human is (reluctantly) opening to incorporate non-normative genders, sexualities and racialised (and less successfully differently-abled) people, the human is once again returned to a universal category under the rubric of climate change, global warming and/or the Anthropocene.

A more articulate challenge to postcolonialism by the discourse of climate change is put forward by Dipesh Chakrabarty. Chakrabarty, author of *Provincialising Europe*, is one of the world's foremost postcolonial scholars whose nuanced challenge needs to be considered carefully. In two recent articles, Dipesh Chakrabarty implicitly asks about the status of the 'anthropos' (Greek for 'human') in the Anthropocene. The first article, 'The Climate of History: Four Theses' (2009), deals with the relation between the human and nature, while the second article, 'Postcolonial Studies and the Challenge of Climate Change' (2012), deals with different conceptions of the human.

In the first article, Chakrabarty (2009) contends that the planetary crisis of climate change demands we collapse the age-old distinction between human and natural history. Human history, human culture has become a geological force and in order to grapple with this, he argues, we need to come to terms with 'deep history' and to reconceive human history as a history *of* a species, one among many others on the planet. For Chakrabarty, this means reconceptualising the struggle for freedom (whether class, gender, slavery or imperial oppression), which he equates to the last 250 years beginning with the Enlightenment, as simultaneously the time when humans switched from wood and renewable fuels to fossil fuels. 'The mansion of freedom,' he writes, 'stands on an ever-expanding base of fossil-fuel use.' For this reason, he argues, we must reconceptualise the narrow project of human freedom in favour of freedom as a struggle not only to preserve human life but also the conditions under which human life and other forms of life can flourish on this planet.

In the second article, Chakrabarty (2012) argues that climate change challenges postcolonial studies by forcing us to deal with a new 'human'. To summarise, Charkabarty argues that historically there are three images of the human. The first is the Enlightenment notion of the human as a universal category with the capacity to bear rights. The second is the postcolonial-postmodern view that views the human as differentiated according to class, sexuality, gender, history, etc. The second sense acts a corrective to the first since underneath it still relies on an Enlightenment conception of the human. The final and most recent image, which challenges the first two, is the figure of the human in the age of the Anthropocene where humans collectively emerge as a geological force on the planet (ibid.: 2).

First, even presuming the periodisation to be uncontested,[6] this dating of the Anthropocene as beginning in the Enlightenment begs the question of its relationship to colonisation and imperialism. The Industrial Revolution, which is the beginning of the ever expanding use of fossil fuels, is praised by historians of modernity as a sign of progress, inventiveness and modern civilisation, while downplaying the centrality of extensive resource extraction from British and other European colonies which actually drove this revolution. Yet the same period viewed as the beginning of the demise of our planetary ecology is decoupled from its European moorings to implicate all humanity under the title 'Anthropocene'. If we take the periodisation at face value, a fairer and more precise term would be Eurocene or Anglocene.

Second, and leaving aside my disagreement with the second image of the human (I do not think it necessary to pose a universal human underneath anthropological difference to maintain a postcolonial position), what is clear here is that the 'challenge' articulated by climate change is primarily a conceptual one. The challenge is double here: the human conceptualised as differently universal and the human reconceptualised as one among many species.

Regarding the first challenge, while I agree to the extent that one can ideally separate these varying conceptions of the human, I am concerned that this separation is never constant in practice. As Chakrabarty himself points out in the second article, these images continue to be in competition with each other in the present (Chakrabarty, 2012: 2). I suggest that this competition can work through modes of co-optation, where the narrow 'human' of the European Enlightenment has the potential to be disguised in Anthropocenic clothes. In relation to the latter challenge, it is difficult to imagine how the conception of the human can be disentangled from nature, given the continued pervasiveness of anthropocentrism and anthropomorphism. In such instances, nature is not only conceptualised in binary terms from the human, it is often humanised and personified in various ways. Consequently, how we deal with nature is often a reflection of how we deal with humans and is thus reflective of human politics. In the next section, I turn to invasion rhetoric of environmental security as one area in which these two problems are apparent.

Imperial ecology and invasion biology

Nowhere is this more visible than in the fact that the fantasy of invasion is a central trope for managing both non-white peoples and plants and animals. Here the sense of more-than-human is always already doubled, coupling the non-human (fauna and flora) with the infra-human (non-European descendants). We see this in, for example, the Australian Research Council's National Research Priority 'Safeguarding Australia', which is described as 'Safeguarding Australia from terrorism, crime, invasive diseases and pests'. Recalling the wider use of viral imagery to portray criminality and terrorism, this fortress imagination of Australia is based on the mixing of terrorism and criminality with pests and diseases, which are conceptualised as simply different threats operating under the same security logic. Both crime and disease are major sites for racially divergent immigration processes in Western border security regimes, which ask questions about the criminal and medical histories of the applicants. Here, 'invasion' rhetoric naturalises the security logics of managing non-Western populations with reference to disease control.

Coming from the other side of the divide, 'invasion' rhetoric in biology can also borrow from human forms of discrimination (specifically racism). Recently, a group of ecologists signed their protest against the discrimination marshalled under invasive species rhetoric. Entitled 'Don't judge species on their origins', a pun on Charles Darwin's classic, *On the Origin of Species*, the 19 scientists claim there is a 'pervasive bias against alien species' across science, government, policy-makers, conservationists and the public in which 'non-native' species are vilified for driving beloved 'native' species to extinction' or 'generally polluting "natural" environments' (Davis *et al.*, 2011: 153). They point out that 'invasion biology' gained prominence in the 1990s, often deploying militaristic metaphors and exaggerated crises to 'convey the message that introduced species are the enemies of man and nature' (ibid.: 153). While there are indeed instances of introduced species becoming threats to the local environment, they continue, 'recent analyses suggest that invaders do not represent a major extinction threat to most species in most environments' and 'in fact . . . non-native species have almost always increased the number of species in a region' (ibid.: 153). They conclude that classifying biota 'according to their adherence to cultural standards of belonging, citizenship, fair play and morality does not advance our understanding of ecology' (ibid.: 153).

The imagined 'natural' ecological system that is being protected is often historically specific. Policy-makers in settler colonial countries do not imagine recreating the environment to the way it was *prior* to colonisation but rather to maintain the status quo. Significant attention is paid to protecting core agricultural products of the economy, which more often than not are non-native species introduced via histories of colonisation. In other words, invasive biology rhetoric often takes for granted the 'naturalness' of human colonisation, even as it claims to be concerned about the 'unnaturalness' of non-human colonisation.

The fact remains that the most invasive species, the one that has wrought large-scale destruction of diverse habitats of millions of other species, the predator that has asymmetrically controlled its prey for thousands of years is in fact humans. But of course when scientists and governments complain about invasive pests, they never once consider humans to be the issue. Indeed, humans who have questioned scientists' and governments' right to destroy natural environments and experiment on other non-human animals are labelled 'eco-terrorists'.

In part, this has to do with the historical trajectories within which ecology emerged. Ecology and conservationism are themselves artefacts of empire, often inspired by colonial settlements (see Griffiths and Robin, 1997). Colonial islands were central environments through which new conceptions of nature were developed, paving the way for conservationist criticisms against the Empire (Grove, 1995). But the rhetoric of preservationism was also used to racially discriminate, such as the distinction between white 'hunting' and black 'poaching' in colonial Africa (McKenzie, 1997).

The terms 'native' and 'alien' were transposed from British common law to botany in the 1840s to distinguish 'true' British flora from others (Davis *et al.*, 2011: 153). In other words, the language of British migration, a system for managing racial populations through the nation-state, becomes the primary language for imagining flora. The growth of ecology from 1895 drew on the machinations of British imperial administrative and political culture, funded by colonial governments and companies to conduct various surveys and expeditions (see Anker, 2001, particularly Chapter 3).

By the time Charles Elton published *The Ecology of Invasion by Animals and Plants* in 1958, in Britain, a seminal text of invasion biology, post-war England was undergoing its own internal debates about mass non-white post-war migration to the heart of the Empire. Indeed, Elton's work was heavily influenced by his teacher at Oxford, Sir Alexander Carr-Saunders, a British biologist and sociologist, who was at one time secretary of the British Eugenics Education Society. In particular, Elton was influenced by Carr-Saunders' (1922) book, *The Population Problem*, a neo-Malthusian work which contended that the population problem was largely due to primitive peoples reproducing at higher rates because of lower mental and physical capacity, which in turn endangered the standard of living among the higher races (and for which he was appointed to a Chair in Social Science at University of Liverpool). As Peder Anker argues in his brilliant study of imperial ecology, many of the ideas central to Carr-Saunders' book reappear in Elton's schema (Anker, 2001: 93–94, 101–102).

To be clear, I am not suggesting that Elton transported wholesale Carr-Saunders' eugenicist politics simply because he borrowed certain concepts. However, precisely because the conceptual framework and language are similar, Elton's invasion biology lends itself to analogies with human systems of managing racial populations. Such analogies are not simply linguistic but carry material weight when applied in systems of control. This is exemplified in the continued use of invasion narratives to manage both animals and humans in contemporary Australian border security regimes, to which I now turn.

The birds and the bees: a contemporary parable of invasive ecologies

In 2005, the reported cases of avian influenza virus H5N1 in Asia initiated yet another media panic about Australia's unprotected shores. *The Australian* warned that 'migrating birds from Southeast Asia could bring the deadly bird flu to Australia' (Wilson and Powell, 2005: 3). The then Health Minister, Tony Abbott, dismissed the risk of poultry infection as low because 'our flocks are kept under very different conditions' in comparison to Southeast Asia and China (cited in Wilson and Powell). By 2009, after further reported outbreaks of bird flu in Asia, the Howard Government invested $114 million to purchase 2 million treatments of Tamiflu to counter the possibility of a bird flu epidemic (even though the Howard Government had destroyed 1.6 million Tamiflu doses that year, which it had bought in 2004 to counter the media panic about a potential swine flu outbreak).[7]

Jeff Ironside, Chairman of the Australian Egg Corporation, was quick to use the panic to attack the smaller, free range poultry farms since, as he argued, large farm organisations had stricter biosecurity measures. However, Earl Brown, a virologist at the University of Ottowa, suggested that it was in fact larger farms that were more at risk: 'High-intensity chicken rearing is a perfect environment for generating virulent avian flu viruses.' Indeed, the 2005 UN Taskforce on the bird flu concluded that one of the root causes was 'farming methods which crowd huge numbers of animals in a small space', in other words, factory farming. As Brown pointed out, viruses in wild birds tend not to be dangerous whereas viruses entering high-density animal populations (e.g. poultry factory farms) provide ample opportunity for the virus to mutate into something more dangerous, especially when factory farming has reduced the genetic diversity of their animals through selective breeding. In other words, precisely the kinds of factory farming practices (intensive livestock operations based on selective breeding and indoor rearing) developed initially by the West during the Agricultural Revolution and further intensified in the Industrial Revolution (see Overton, 1996: 111). Such intensive farming practices were subsequently transported to its colonies and later to 'developing' (read: postcolonial) countries as part of international aid programmes to stimulate the modernisation of the non-West.

This history, however, is obfuscated by these kinds of panics that tend to paint the non-West as a breeding ground for viruses, a ground zero for disease outbreaks. Film serves as a significant cultural vehicle for circulating this geopolitical epidemic imaginary, by which I mean how the globe is imagined in terms of viral outbreaks and risks. Epidemic panics become inspirations for films like *Contagion* (2011), in which the US Centers for Disease Control feature as the hero, saving the world from the outbreak of a mutant swine flu virus. In its closing flashback sequence, the film pins the virus on Chinese deforestation and the unsanitary practices of Chinese farming and food handling.

Contagion joins a list of epidemic disaster films and television shows since the 1970s – films like *Outbreak* (1995) and *Pandemic* (2009) as well as television series such as *Fatal Contact: Bird Flu in America* (2006) and *Pandemic* (2007) – in which various Third World countries are blamed for fatal disease outbreaks, which, thanks to the tireless efforts of American scientists, are deflected from destroying the entire human population.[8] Such cinematic imaginations of epidemics visualise Third World countries as dangerous viral breeding grounds of ignorant masses, which ultimately function as symptoms of their morally and politically corrupt governments. These continue a much longer history that associates non-normative racial, sexual and gendered minorities and the working class with disease and disorder (see, for example, Bashford, 1998, 2004; Peckham and Pomfret, 2013).

Meanwhile, entomologists are busy trying to protect native ecosystems from the onslaught of alien insects. Globalised world trade and mobility translocate alien insect species by human agency at an alarming rate, we are told. Hawaii accumulates 20–30 new insect species per year while 12–15 new species are reported on Guam (Samways, 1999). These alien invasions, we are told, are devastating the character and stability of natural ecosystems, landscapes, regions and islands, which in turn decreases agricultural output (ibid.).

Within Australia, this entomological crisis has centred on bees. Western Australia's Department of Agriculture's website warns of 'The Exotic Bee Threat' because 'bees can carry exotic bee diseases'.[9] In particular, 'Asian bees' and 'African bees', including the 'Africanised bees' of Brazil, are singled out, even though the website informs us that all foreign bees are threats. Apparently, the Asian bee has 'spread to Papua New Guinea' and 'has a high propensity to swarm and will readily swarm and establish nests on ships and cargo'. The website provides a picture contrasting the European, Western or 'Common' honeybee with the Asian bee, which appears smaller and darker. Due to Australia's proximity to Papua New Guinea and other 'Asian regions', its sea ports are particularly prone to the Asian bee danger. In contrast to the Asian bee, the African bee is portrayed as having 'aggressive defence characteristics', and through interbreeding this violent streak has spread to the Brazilian bees, which are thus referred to as 'Africanised bees'. A beekeeper explains that Africanised bees 'just keep on attacking in mass [*sic*] for much longer than European bees would' and explains that a 'disadvantage of Africanised bees is that they mate with European bees' and for this reason Australia does not allow in bees from the US.

The barely hidden parable of Australian migration anxiety is striking. Replace the word 'bee' with 'people' or 'culture' and one ends up with a neatly encapsulated picture of White Australian stereotypes of race in a globalised world. On one side, we have the yellow hordes swarming, or 'swamping' in Pauline Hanson's terms, the nation from the North, which have been recycled in various invasion narratives in Australian literature; and, on the other side, the Africans, often mistakenly referred to as 'Sudanese' in Australia, who are prone to congregate in dangerously violent gangs, as portrayed by mainstream media. Meanwhile, the non-native status of European honey bees has long since been forgotten. The irony that we are meant to protect the European from 'exotic invasion' remains unquestioned. But if this apiarian theatre is a space to restage the naturalisation of European Australians, it also highlights the dependence of this naturalisation on the exoticisation of others. The White Australian, in other words, can only be imagined as native to Australia by continuously invoking the foreignness of others (Asians, Africans, Indians, etc.).

This perpetual invocation of the foreignness of coloured others and the nativeness of the white self, is recaptured in the continuous replay of scenes of suspicious foreigners entering Australia. This scene of the border is played again and again on the hit television program, *Border Security: Australia's Frontline*. *Border Security*, which follows the work of Australian Customs and Border Protection, the Australian Quarantine and Inspection Service and the Department of Immigration and Citizenship as they go about their daily work of enforcing Australian customs, immigration and finance laws and quarantine practices. The program is supported by the government departments and agencies involved, who have the right to veto what gets aired, which introduces a political bias in the post-production process (Burton, 2007). As Deveny (2007) argues, the program is thus a mixture between television documentary, security advertisement, government propaganda and public information campaign.[10] Consequently, the program has been the butt of several critiques by journalists aiming to expose its public relations overtones.

Such bias does little to curb the Australian public's fascination with the program which has regularly appeared in the top ten watched television programs since 2005. According to audience research funded by Australia's Customs and Border Protection Services itself, 78 per cent of Australians thought the program was 'realistic'. The aim of the research was to assess to what extent Customs involvement in *Border Security* produced positive outcomes for the government agency. DMB Consultants, who conducted the research, found that 96 per cent of the 1.2 million viewers thought Customs did an important job and 83 per cent said it 'keeps Australia safe', claiming the program was informative, amusing, realistic and relevant. Yet when asked about the role of Customs, viewers were confused. The program was most popular among people from lower-income brackets who had not travelled in the previous year (constituting 93 per cent of the viewers) and who were more likely to believe that Customs and Border Protection were responsible for 'preventing illegal immigration' (Meade, 2011). From one angle we might say the entertainment of watching is dependent in part on confusing border protection from pests, diseases and weapons with border protection from foreigners, who are seen as bearers of these ills. It may employ different suspicious coloured characters (a Chinese couple visiting their student son, a Muslim Malaysian couple returning to Australia, a black African couple visiting Australia or an elderly woman from Hong Kong visiting her daughter) but the message is the same. At best, the program portrays foreigners as ignorant. At its worst, it portrays coloured foreigners as arrogant and scheming carriers of exotic diseases and pests.

Conclusion

Nature is an image of human making in which humans can reflect on themselves and their society. Consequently, attempts to manage nature are always already invested with human constructs and ultimately therefore open to manipulation in human politics. The deployment of invasion narratives in Australia's border security regime (though the analysis can easily be extended to other countries) reflects this on-going imbrication between the control of nature and control of humans, between ecology and eugenics, between conservationism and protectionism.

It is at this level that I have concerns with the universal human conception at the heart of the Anthropocene because it has the potential to be used to reinstall the white liberal human subject of the Enlightenment. We can already see this in media attempts to locate the cause of climate change in Third World countries' pollution and deforestation practices. These countries are therefore portrayed as the cause of 'our' (read: Western) decline in living standards not simply in the present but importantly in the future. Such arguments echo the racism touted by Carr-Saunders' *The Population Problem*; this time through climate change. This cultural anxiety around the polluting presence of non-Western countries obfuscates not simply the history of colonisation and the British Industrial Revolution that introduced such problems to the world, but also the present massive discrepancy of pollution per capita between OECD countries and developing countries.

Yet the legacy of imperial ecology has also given us science fiction and with it the possibility of imagining this kind of dystopic world. An environmentally challenged world, far from uniting humanity under a universal 'human', might be precisely the rhetorical space needed to reinvigorate old prejudices. There is a strong eschatological literary tradition that has helped us imagine the dangers of a post-human world (see Johnson, 2012). This literary tradition continues in more recent natural disaster films, which envisage a post-apocalyptic world where power is reinscribed through and onto environmental disasters (through the control of water, control of reproduction). It is for this reason that postcolonialism must remain one of the limits that tempers the excesses of the enthusiasm for the Anthropocene in the humanities.

Notes

1 A meeting on the Anthropocene was held at the Geological Society of London on 11 May 2011, which featured in *Nature, Science* and *The Economist* and on the BBC. The Anthropocene was a major theme at the 2011 Nobel Laureate Symposium and the Mineralogical Society of Poland invited a keynote talk on the Anthropocene in September 2011. The *Philosophical Transactions of the Royal Society* (2011) published a special edition devoted to the Anthropocene. Meanwhile, journal articles on the natural sciences on the Anthropocene increased to 331 in 2011 and Jacques Grinvald, an historian of science, is publishing a new edition of his book *La Biosphère de l'Anthropocène* in 2012, which details the history of the concept of the Anthropocene from the biosphere.

2 See http://quaternary.stratigraphy.org/workinggroups/anthropocene/ (accessed 11 February 2013).

3 For work on ecopoetics, see Bryson (2002) and Killingsworth (2004). Ecocriticism is now a large body of literature but for a good introduction, see Garrard (2004). For earlier examples of ecocriticism, see Glotfelty and Fromm (1996) and Kerridge and Sammells (1998), and see Coupe (2000) to trace its links to earlier Romanticism. For earlier environmental criticisms of capitalism, see Caldwell (1972) and Stretton (1976). For

ecological Marxism, see O'Connor (1994) and O'Connor (1998), while for a materialist reinterpretation of environmentalism, see Burkett (1999). I use ecological anarchism broadly to encompass Bookchin's (1971) early attempt to construct a post-scarcity anarchism and, more recently, Carter's (1999) more considered approach to an anarchist environmental politics. Environmental security is now an expansive literature but Dalby's (2002) book is still the best critical work on the topic in security studies and international relations.

4 I understand poststructuralism to be an Anglophonic term that encompasses a variety of approaches influenced by French theorists (such as Derrida and Foucault) who use language-based analyses, but which have differing stances on the relationship between language and reality. I understand anti-foundationalism to make the particular claim that there is no reality outside language and consequently I see this as a specific strand of poststructuralism.

5 This is a staple point of postcolonial and decolonial thought but see, for example, Gilroy (2000) and Goldberg (1993, 2002).

6 There is some disagreement about Crutzen's periodisation of the Anthropocene beginning with the invention of the steam engine in the late eighteenth century (see Smith and Zeder, 2013). On the other hand, Ruddiman (2003, 2005a, 2005b, 2006) champions an early-Anthropocene hypothesis, dating its beginning to mass deforestation 8000 years ago based on atmospheric composition (see also the special edition of *Holocene*, vol. 21, edited by Ruddiman, Crucifix and Oldfield in 2011), while Dull *et al.* concur, suggesting deforestation peaked in the late-fifteenth century when European arrival in the New World brought new diseases which resulted in a 95 per cent reduction of the Indigenous populations.

7 The naming of diseases is a fascinating reflection of geopolitics. US Secretary of Agriculture Tom Vislak expressed concerns that the term 'swine flu' would convince people not to eat pork and consequently the US Centers for Disease Control re-named it the 'Novel Influenza A (H1N1)', which was picked up by the European Commission's use of 'novel flu virus'. Meanwhile, in the Netherlands, it was originally known as 'pig flu' but eventually renamed the 'Mexico flu' by the Netherlands National Institute for Public Health and the Environment, which is similar to 'Mexican virus' adopted by both South Korea and Israel. The World Organisation for Animal Health proposed 'North American influenza', which unsurprisingly was not taken up.

8 I should note that the zombie film genre appears as the political counter-point to the epidemic disaster film, since it is usually the moral decay of the West (either governments, scientists or capitalist corporations) that serves as the breeding ground for the zombie virus.

9 See http://agspsrv34.agric.wa.gov.au/ento/bee3.htm (accessed 18 November 2012). All other references in this paragraph refer to this webpage.

10 For a broader discussion of the rise of 'securitainment' in the post-9/11 environment, see Andrejevic (2011).

References

Adams, C. J. 1990. *The Sexual Politics of Meat: A Feminist-Vegetarian Critical Theory*, Oxford, Polity Press.

Andrejevic, M. 2011. 'Securitainment' in the Post-9/11 Era. *Continuum: A Journal Of Media and Popular Culture*, 25, 165–175.

Anker, P. 2001. *Imperial Ecology: Environmental Order in the British Empire, 1895–1945*, Cambridge, MA, Harvard University Press.

Anthropocene Working Group. 2009. Anthropocene Working Group of the Subcommission on Quarternary Stratigraphy (International Commission on Stratigraphy), *Newsletter* 1,

December 2009. Available at: http://www.quaternary.stratigraphy.org.uk/working groups/anthropocene/ (accessed 6 March 2012).

Bashford, A. 1998. *Purity and Pollution: Gender, Embodiment and Victorian Medicine*, New York, Macmillan.

Bashford, A. 2004. *Imperial Hygiene: A Critical History of Colonialism, Nationalism and Public Health*, London, Palgrave Macmillan.

Bookchin, M. 1971. *Post-Scarcity Anarchism*, Berkeley, CA, Ramparts Press.

Bryson, J. S. 2002. *Ecopoetry: A Critical Introduction*, Salt Lake City, UT, University of Utah Press.

Burkett, P. 1999. *Marx and Nature: A Red and Green Perspective*, Basingstoke, Macmillan.

Burton, B. 2007. Need to Curb a PR Industry Spinning Out of Control. *Sydney Morning Herald*, 7 August 2007. Available at: http://www.smh.com.au/news/opinion/need-to-curb-a-pr-industry-spinning-out-of-control/2007/08/06/1186252624551.html (accessed 14 June 2012).

Caldwell, M. 1972. *Socialism and the Environment*, Nottingham, Bertrand Russell Peace Foundation.

Carr-Saunders, A. M. 1922. *The Population Problem: A Study in Human Evolution*, Oxford, Clarendon Press.

Carter, A. 1999. *A Radical Green Political Theory*, London, Routledge.

Chakrabarty, D. 2009. The Climate of History: Four Theses. *Critical Inquiry*, 35, 197–222.

Chakrabarty, D. 2012. Postcolonial Studies and the Challenge of Climate Change. *New Literary History*, 43, 1–18.

Coupe, L. 2000. *The Green Studies Reader: From Romanticism to Ecocriticism*, London, Routledge.

Dalby, S. 2002. *Environmental Security*, Minneapolis, MN: University of Minnesota Press.

Davis, M. A., Chew, M. K., Hobbs, R. J., Lugo, A. L. E., Ewel, J. J., Vermeij, G. J., *et al.* 2011. Don't Judge Species on Their Origin. *Nature*, 474: 153–154.

Deveny, C. 2007. Couch Life: Heading the Bogans off at the Pass. *The Age*, 18 August.

Dull, R. A., Nevle, R. J., Woods, W. I., Bird, D. K., Avnery, S. and Denevan, W. M. 2010. The Columbian Encounter and the Little Ice Age: Abrupt Land Use Change, Fire, and Greenhouse Forcing. *Annals of the Association of American Geographers*, 10, 4, 755–771.

Elton, C. 1958. *The Ecology of Invasions by Animals and Plants*, London, Methuen.

Garrard, G. 2004. *Ecocriticism*, London, Routledge.

Gilroy, P. 2000. *Against Race: Imagining Political Culture Beyond the Color Line*, Cambridge, MA, Belknap Press of Harvard.

Glotfelty, C. and Fromm, H. (eds) 1996. *The Ecocriticism Reader: Landmarks in Literary Ecology*, Athens, GA, University of Georgia Press.

Goldberg, D. T. 1993. *Racist Culture: Philosophy and the Politics of Meaning*, Cambridge, MA, Blackwell.

Goldberg, D. T. 2002. *The Racial State*, Malden, MA, Blackwell.

Griffiths, T. and Robin, L. 1997. *Ecology and Empire: Environmental History of Settler Societies*, Edinburgh, Keele University Press.

Grove, R. 1995. *Green Imperialism: Colonial Expansion, Tropical Island Edens and the Origins of Environmentalism, 1600–1860*, Cambridge, Cambridge University Press.

Grove, R. 1997. *Ecology, Climate and Empire: Colonialism and Global Environmental History, 1400–1940*, Cambridge, White Horse Press.

Hodson, M. and Marvin, S. 2010. Urbanism in the Anthropocene: Ecological Urbanism or Premium Ecological Enclaves. *City: Analysis of Urban Trends, Culture, Theory, Policy, Action*, 14, 298–313.

Johnson, K. L. 2012. Introduction to Focus: Exanthropic Poetics. *American Book Review*, 33, 3.

Kerridge, R. and Sammells, N. (eds) 1998. *Writing the Environment: Ecocriticism and Literature*, London, Zed Books.

Killingsworth, M. J. 2004. *Walt Whitman and the Earth: A Study in Ecopoetics*, Iowa, University of Iowa Press.

McKenzie, J. M. 1997. *The Empire of Nature: Hunting, Conservation and British Imperialism*, Manchester, Manchester University Press.

Meade, A. 2011. Customs Sniffs out a Winner with Reality TV. *The Australian*, 2 May. Available at: http://www.theaustralian.com.au/media/customs-sniffs-out-a-winner-with-reality-tv/story-e6frg996-1226047993404 (accessed 12 February 2012).

O'Connor, J. 1998. *Natural Causes: Essays in Ecological Marxism*, New York, Guilford Press.

O'Connor, M. (ed.) 1994. *Is Capitalism Sustainable? Political Economy and the Politics of Ecology*, New York, Guilford Press.

Overton, M. 1996. *Agricultural Revolution in England: The Transformation of the Agrarian Economy 1500–1850*, Cambridge, Cambridge University Press.

Peckham, R. and Pomfret, D. M. 2013. *Imperial Contagions: Medicine, Hygiene and Cultures of Planning in Asia*, Hong Kong, Hong Kong University Press.

Phelan, L., Henderson-Sellers, A. and Taplin, R. 2012. The Political Economy of Addressing the Climate Crisis in the Earth System: Undermining Perverse Resilience. *New Political Economy*, 18, 198–226.

Ruddiman, W. F. 2003. The Anthropogenic Greenhouse Era Began Thousands of Years Ago. *Climatic Change*, 61, 261–293.

Ruddiman, W. F. 2005a. How Did Humans First Alter Global Climate? *Scientific American*, 292, 46–53.

Ruddiman, W. F. 2005b. The Early Anthropogenic Hypothesis A Year Later (and Editorial Reply). *Climatic Change*, 69, 427–434.

Ruddiman, W. F. 2006. Have Humans Altered Global Climate for Thousands of Years? *Geoscope*, 37, 1–5.

Samways, M. J. 1999. Translocating Fauna to Foreign Lands: Here Comes the Homogenocene. *Journal of Insect Conservation*, 3, 65–66.

Smith, B. and Zeder, M. 2013. The Onset of the Anthropocene. *Anthropocene*. Available at: http://dx.doi.org/10.1016/j.ancene.2013.05.001 (accessed 12 June 2013).

Stretton, H. 1976. *Capitalism, Socialism and the Environment*, Cambridge, Cambridge University Press.

Wilson, A. and Powell, S. 2005. Local Flu Risk from Migrating Wild Birds. *The Australian*, 24 September, 2005, p. 3.

Films

Connery, J. director, 2009. *Pandemic*, Alianza Films International.

Petersen, W. director, 1995. *Outbreak*, Warner Bros.

Soderbergh, S. director, 2011. *Contagion*, Warner Bros.

Television

Mastroianni, A. 2007. *Pandemic*, Hallmark Entertainment.

Pearce, R. 2006. *Fatal Contact: Bird Flu in America*, ABC.

4

RESILIENCE IN THE ANTHROPOCENE

A biography

Libby Robin

Anthropocene anxiety in the new millennium

As we entered the present millennium, there was a new anxiety, something much bigger than a mere calendar moment, something more than the growing concern about global warming which had been gathering pace since the greenhouse summer of 1988 (Hulme, 2013). Paul Crutzen, a Nobel Prize-winning atmospheric chemist, felt it keenly. He wanted to capture the collective insights of the physical and biological Earth sciences that indicated that the planet was changing fast in all sorts of ways. All of the changes were being induced or 'forced' by human activities. Crutzen proposed that the Earth had entered a new geological epoch, the Anthropocene, where humanity is now a global force changing biological and physical Earth systems (Crutzen and Stoermer, 2000).

Since his initial proposal, Crutzen's Anthropocene has figured prominently in discussions of environmental change. There are three broad areas of debate: the first is in the geological and stratigraphic science community about whether the changes justify the end of the Holocene epoch, the last of the current series, and the beginning of a new one. Geological epochs are usually indicated by stratigraphic layers in rocks, but the Anthropocene is as much about the acidity of oceans, the presence of greenhouse gases in the air and changing ecologies, so it requires technical ways to establish equivalence with earlier epochs (Zalasiewicz *et al.*, 2011). The second area of Anthropocene debate is around the crucial historical question: *when did it begin?* (Robin, 2013). The third area is raised by the adoption of the Anthropocene as an heuristic device. Scholars of many persuasions well beyond geology are considering anew the place of humanity on earth. From the notion that humans are themselves invasive species, to the question of how humans might adapt to planetary responses to historical changes, the Anthropocene gathers together a range of new experts and new planetary discourses.

Leading figures in the Anthropocene discourse agree on a small number of indicators to show that human beings have already exceeded the planetary boundaries that frame 'a safe living space' on Earth (Rockström *et al.*, 2009). Of the nine planetary boundaries measured, the greatest concern was about extinctions. The loss of species to date – compared with the natural attrition in an evolutionary curve – corresponds to several planets' worth of evolutionary change. Thus the causes of extinctions, particularly human-induced causes, are a major centre of Anthropocene activity.

Oxford ecologist Charles Elton ran a popular series of radio broadcasts during the 1950s decade of nuclear testing. They were all about explosions. He explained:

> An ecological explosion means the enormous increase in numbers of some kind of living organism – it may be an infectious virus like influenza, or a bacterium like bubonic plague, or a fungus like that of the potato disease, a green plant like the prickly pear, or an animal like the grey squirrel. I use the word 'explosion' deliberately, because it means the bursting out from control of forces that were previously held in restraint.
>
> *(Elton, 1958: 15)*

His book, *The Ecology of Invasions by Animals and Plants*, positioned widely distributed animals and plants as *invaders*. Elton also described these organisms as part of the baggage of humans as they moved around the world. Ecologists, grappling with the increasing prevalence of invasive species, one of the major indicators of what is called the Great Acceleration in Anthropocene discourse, seized on Elton's metaphor and described their work as *invasion biology*, celebrating its jubilee in 2008, just as the drivers of Anthropocene change were reaching criticality (Richardson, 2011; Steffen *et al.*, 2011).

Another response to the Anthropocene crisis was the Stockholm Resilience Centre (SRC, http://www.stockholmresilience.org/), established in 2007, an institution with a brief to research the governance of social-ecological systems, with a special emphasis on resilience. The SRC uses the ecological concept of resilience to include social systems and focus on the ability to deal with change and continue to develop. Johan Rockström, lead author of the planetary boundaries paper mentioned above, and Carl Folke, an ecologist with strong credentials in economics, are the directors of the SRC. Resilience is thus one of the prominent concepts in the understanding of the Anthropocene.

Resilience is a powerful concept when extended from its origins in ecology to discussions of natures, cultures and societies in the era of the Anthropocene. Claude Lévi-Strauss commented that: 'Words are instruments that people are free to adapt to any use, provided they make clear their intentions' (Braudel, 1995: 3). This chapter reviews the emergence and redefinitions of the word and the concept of resilience, providing some historical context for current debates. It is now both a metaphor and a conceptual tool that transcends its origins in ecology.

C. S. Holling and resilience

Resilience was defined as an ecological concept in 1973 by C. S. (Buzz) Holling, an animal ecologist and evolutionary biologist interested in mathematical models of predator–prey relations. His 1973 definition, still frequently cited, was that resilience was a measure of 'the persistence of systems and their ability to absorb change and disturbance' (Holling, 1973: 14). Holling contrasted resilience with *stability*, 'the ability of a system to return to an equilibrium state after a temporary disturbance' (ibid.: 14) The value of the concept of resilience in ecology was that it described the system as it responded to a changing world whether there was equilibrium or not. It theorised the relationship between the shocks and the system, and was suitable for assessing the effects of the activities of people on natural systems.

A resilient system keeps absorbing successive shocks and adapting without dramatically altering the relationships between its parts. But at some point this resilient character passes a limit and beyond this, 'the system rapidly changes to another condition' (ibid.: 7). The example Holling gave was of a small Italian lake whose limnology was unchanged for a very long period, even though the catchment around it changed from steppe to grassland to forest. Then, suddenly, the hydrography of the lake changed and it became eutrophied (rich in nutrients). What had changed? The eutrophication corresponded to the building of the Roman road, the Via Cassia, about 171 BC. The surrounding landscapes could have very little cover or whole forests without changing what could or could not live in the lake. But a hard Roman road and people using it suddenly altered the way water flowed into the lake. It was the anthropogenic change that altered and decided which animals and plants could survive there.

Resilience is the stretch of the system, what it can absorb and adapt to, but it is not an infinite property. It has *limits* (ibid.: 7). Limits were very much in vogue in the 1970s. The famous report, *Limits to Growth* (Meadows et al., 1972) was published the year before this paper. The language has changed. In the 1970s and 1980s, people spoke of 'abrupt change' or 'flips' in the system (Broecker, 1987: 125). Now, more commonly, we hear terms like 'threshold levels', 'tipping points' or 'planetary boundaries' (Lindsay and Zhang, 2005; Rockström et al., 2009, Bhatanacharoen et al., 2011). The scale is more often global, but the idea of focusing on the system (even the whole Earth) under stress, has gathered pace since Holling's original resilience paper. Resilience enabled ecologists to think about which properties of a system maintained its *essence*, and which forced a system to shift from one state to another.

Ecology has moved away from the idea of natural balance. Resilience is one of the concepts it has developed to describe interactions between dynamic natural systems, and increasingly between natural and social systems (Egerton, 1973). Resilience is also a key metaphor in business and medicine, particularly psychiatry, the study of the psyche under stress (Luthar, 2006). Indeed, it sometimes works near the nodes of *connections* between ecology, business and medicine (sustainable ecosystems; the health of the planet; the breakdown of systems). This particular theory of ecology has become an attractive metaphor enabling transdisciplinary dialogue, and perhaps has encouraged a proliferation of medical metaphors in environmental management.

Through tracing the biography of an idea, and in paying attention to the history and geography of resilience, it is possible to find geographical origins even in global concepts circulating through the Anthropocene literature. Ecology began with specific local ecosystems. Historically, the closed, finite ecosystems of North American freshwater lakes and their fisheries were carefully documented. In the late nineteenth century, Stephen Forbes, later President of the Ecological Society of America, spoke of the freshwater lakes as 'microcosms' for study. His Illinois lake was a 'little world within itself, a microcosm within which all the elemental forces are at work and the play of life goes on in full, but on so small a scale as to bring it easily within the mental grasp' (Forbes, [1887] 1925: 537; Golley, 1993: 36–37; Schneider, 2000). The relationship between elements was relatively simple and could be described by linear relationships because the lake was 'climatically buffered, fairly homogenous and self-contained' (Holling, 1973: 18). But the rapidly changing world of the latter half of the twentieth century, where cities were growing, land-clearance rates were high and pollution was changing the way the natural world worked, demanded new models. Even the heterogeneous (patchy) landscapes of the arid rangelands in the western United States and Australia did not function as closed systems. They were prey to invasive species, and consequent hydrological change, and different parts of the landscape responded differently as land use changed. The idea of assessing the ecosystems as if they were stable made little sense. The resilience framework suggested a different management strategy, which de-emphasised the predictable world, and favoured keeping options open 'to absorb and accommodate future events in whatever unexpected form they took' (ibid.: 21). It was a model that recognised that change, not stability, was defining ecosystems.

Science from somewhere

Science traditionally aspires to universal truths, that is, concepts that apply whenever the preconditions specified are met. Yet the science of ecology has developed as a very specific science of place. Since Alexander von Humboldt's nineteenth-century travels in search of climatic parallels, biogeographies of climatic and latitudinal similarity have invited comparison. Even so, ecological systems remain dependent on edaphic (soil or geological) conditions, on temperature and elevation, on 'edge effects' and increasingly on encroachments from human society, including massive shifts of biota from continent to continent. In short, there is a veritable suite of preconditions making it difficult to transfer ecological knowledge of one place to another without the interpretation of a *locally* expert ecologist. Raymond Specht, the ecologist who co-ordinated Australia's national contributions to the International Biosphere Program (IBP) in the 1970s, commented that ecological facts were too context-dependent to be mapped directly into any computer mega-system without considerable mediation by ecologists with knowledge of the ecosystems in question (Specht, interview with the author, 1991). Such an admission was an embarrassment for an international programme. The historian of science, Steven Shapin has described this paradox: 'The claim that [scientific] knowledge is geographically located is widely taken as a way of saying that the knowledge in question is not authentically true at all' (Shapin, 1998: 8).

Ecologists frequently described their science as 'young'. It had yet to grow up. If it was just local and place-dependent, it was not proper science at all (Cittadino, 1980). The ecological concept of resilience has aided ecology in its (at times) painful move from local field science to force in global change thinking. Its history reveals that ecology has not, in fact, shed its place-dependence, but rather developed ways to scale its local from single ecosystems to interconnected ecosystems to planetary systems. In the process of up-scaling, ecological science has increasingly embraced the human as well as the non-human (natural) world in its scope of influence. Some practitioners have used resilience to reinforce management techniques and related ideas such as crisis policy-making, likening their work to that of a hospital emergency ward (Soulé, 1985).

Science is locally produced, but it often travels with great efficiency. Geographer David Livingstone (2003) has argued that scientific ideas do not develop in the air, but rather through particular geographies, and in the relations between them. Shapin (1998) also urges us to understand not only how knowledge is made in specific places, but also how transactions occur *between* places. A *view from somewhere* can become true or global at a time when places and ecologies have increasingly interpenetrated each other's domains, as we live in the Anthropocene era where the biophysical force of the human species can be felt in the outer atmosphere and the depths of the ocean. Even global ideas have roots.

The desert ecological models that described the variability of climate and measured resources and their uncertainties, contributed a vision that has been stretched to conceptualise the planet. Earth itself is now under the unpredictable, patchy, unprecedented force of climate change. It is witnessing massive extinctions. Some say we are in the grip of the planet's 'sixth mass extinction' event (Leakey and Lewin, 1996). As global change increases variability and unpredictability everywhere, the way people, animals and plants have adapted to living in a variable and unpredictable place like outback Australia, has come to provide a model for imagining planetary futures.

Invasion biology

Biological invasions might be considered permanent interruptions to a natural order, or an unnatural new order. In the 'explosive' world of Charles Elton, ecological explosions are slower than nuclear ones, but 'can be very impressive in their effects' (Elton, 1958: 15). Elton observed that such population explosions usually involve human vectors and affect people. In the 1919 influenza pandemic, microbes that lived within people were the invasive species. Ecological explosions have a global span, but the same invasions may appear at different times in different places. For example, an 'explosion' of infected rats created the bubonic plague in Europe in the twelfth and thirteenth centuries, moving slowly north, to infect Iceland in 1402 (Hastrup, 2009). The Bubonic Plague (or Black Death) emerged again in Manchuria (China) as late as 1911. Predicting when and where a minor fluctuation in populations will drive an invasive explosion is very difficult (Elton, 1958). Although Elton mentioned ecological explosions driven by shifts in native populations, his focus was primarily on outbreaks 'that occur because a foreign species successfully invades another country', often with assistance from human movements' (ibid.: 18).

Elton's work anticipated historian Alfred Crosby's *Ecological Imperialism* (1986) by three decades. Crosby's thesis was that the extraordinary expansion of European peoples over the past five centuries was only possible because of the *avalanche* of the biota that travelled with the immigrants. Europeans arrived in the New World were accompanied by 'a grunting, lowing, neighing, crowing, chirping, snarling, buzzing, self-replicating and world-altering avalanche', Crosby wrote (ibid.: 94). The 'portmanteau biota' gave conquering Europeans a strategic advantage.

While Crosby's theory of ecological imperialism appeals to historians interested in cause and effect, ecologists and environmental managers focused on the invasive species preferred Elton's rhetoric as a framework for ideas about managing invasions (Richardson, 2011). In the 1980s, the Scientific Committee on Problems of the Environment (SCOPE) launched a worldwide assessment of the ecology of bio-logical invasions under the auspices of the United Nations group, the International Council of Scientific Unions (ICSU). SCOPE asked questions amenable to research by the scientific method, such as: (1) What biological characteristics make an invader?; (2) What makes a natural ecosystem susceptible to invasion?; (3) How do we predict (quantitatively) the outcome of a particular introduction?; and (4) What is best practice for managing and conserving natural and semi-natural ecosystems? (Huenneke *et al.*, 1988: 8–9).

The so-called normal functioning of a dynamic ecosystem was central here; this was something that the biological invasion disrupted, and that responded to or adapted to the invasion. The disruptions can, of course, lead to total change, even collapse of the ecosystem in question. So the capacity of the system to recover, to be resilient, in the face of invasion or change, is a key question in invasion biology at all scales from the local to the international. Elton's imaginative leap was to conceptualise biota as *invaders*, to give them agency, and to construct them as a worthy enemy to be managed. Successive generations have taken up the fight against invasive species in the interests of biological diversity.

Defining resilience and the new domain of global change science

Resilience is a scientific concept that has been applied to policy, economics and global change in its broadest sense (quite apart from its use in psychiatry). Although it originated in the science of ecology as a way to describe whole-landscape response to environmental change, it now reaches beyond specific local biophysical systems, and is used to describe global change in social-ecological systems (SES) (Olsson *et al.*, 2004; Walker *et al.*, 2004). The shift from the application to bio-physical systems to those that explicitly embrace politico-cultural elements (including economics, demographics and sociological structures) poses problems that scientists active in this field debate among themselves and at times contest vigorously (Carpenter *et al.*, 2001).

Resilience has a plain English meaning, outside the boundaries of the scientific community, and this is both an asset and a liability for the technical term. It has come to mean 'an ability to confront adversity and still find hope and meaning in life', according to Australian social commentator Anne Deveson. But as Harvard psychiatrist George Vaillant remarked wryly: 'We all know perfectly well what resilience means until we try to define it' (Deveson, 2003: 6).

Resilience is most commonly recognised in the language of the psychology of stress, and is often discussed in relation to soldiers returning from warfare. Being resilient is a positive adaptive response, an alternative to becoming a victim of the changed circumstances. (Goodall and Cadzow, 2009: 4). The definition of resilience as 'flexibility over the long term' and the ability to adapt to changed circumstances is applied widely, for example, in architecture and urban planning (Pickett *et al.*, 2004).

The high positive moral value of resilience has made it attractive to a range of scientists beyond ecology, particularly those seeking to work closely with policy-makers. The cost, however, is that the word can lose its precision for scientific purposes. The general meaning can stretch resilience too far, rendering it a panchreston, a notion that is good for everything because it merely means whatever people want it to mean (Lindenmayer and Fischer, 2007). The scientific resilience community has consistently tried to define it precisely, and to resist the 'anything goes' definition, adopting a range of strategies, including intense internal debates, which continue to the present.

The moral import of resilience has been explicitly debated. 'Unlike sustainability, resilience can be desirable or undesirable', ecologist Steve Carpenter and his colleagues commented (Carpenter *et al.*, 2001: 766). Ecologists endeavoured to move it outside the realm of metaphor where the experts were literary, and into the world of measurement, more familiar to science. Although metaphor was useful because it had transdisciplinary appeal, measurement was critical to a scientific team whose expertise largely rested on numbers. Numbers are also preferred by policy-makers because of their perceived objectivity (Porter, 1995). Carpenter, who works on the zoology of North American lake systems, defined resilience as a precision measuring tool, part of credible science, preferring to leave the moral decisions to policy-makers with expertise in sustainability, another 'catch-all' word. Carl Folke, scientific director of SRC, was less worried about fostering an alliance with sustainability. He argued that resilience was a new way to 'operationalise sustainability', a 'perspective' for analysing social-ecological systems (Folke, 2006).

Both Carpenter and Folke work within the resilience science community, where it is generally agreed that resilience is a tool used to measure complex adaptive systems, but even they have different takes on what exactly should be its moral import for 'the complex adaptive systems' under scrutiny. Typically there is more contestation about the definition at the social end of the spectrum than at the biophysical. Resilience is most contested where policy is implicated. A careful analysis of the internalist resilience literature by Fridolin Simon Brand and Kurt Jax has revealed that resilience has been used descriptively, as a hybrid (descriptive-normative) and as a completely normative term by various practitioners (Brand and Jax, 2007: Table 1).

Despite these differences in how to apply the term, there is surprising unanimity among writers in this community about the *history* of the idea of resilience. Most writers start their discussions of the concept of resilience with the 1973 paper in the *Annual Review of Ecology and Systematics* by Canadian ecologist Crawford Stanley (Buzz) Holling (1930–) 'Resilience and Stability of Ecological Systems' (Holling, 1973). This is remarkable, as papers from the 1970s are rarely cited in current scientific literature, yet this paper continues to be cited prominently, often as definitive of resilience, from that time to the present despite the fact that the concept itself has been considerably nuanced since 1973.

Perhaps more important than the definition of a concept, Buzz Holling's paper of 1973 created a beginning for a community of interdisciplinary global change thinkers working between policy and science. Holling was explicit about the different management strategies needed for stable and heterogeneous systems. The linear models arising from the closed lake systems studied by traditional ecologists like Forbes had less to offer managers working with the complex social systems of modernity (and with people and environments together) than the more heterogeneous and unpredictable systems of rangelands, where the spatial mosaics and demands of making a livelihood introduced 'surprises' into the system. The resilience community defined itself through applying the ecological concept of resilience to the systems it studied. As they looked for surprises, rather than stability, the human dimensions of climate change and correlated social changes emerged more markedly. This was an application on a planetary scale, but it still had recognisable roots in disturbed ecosystems, and in planning with human activities as variables. Resilience came of age in the twenty-first century with the first international resilience conference in Stockholm in April 2008, hosted by the Stockholm Resilience Centre (http://resilience2008.org/resilience/?page=php/main). The second resilience conference was held in Tempe, Arizona, in March 2011 (http://rs.resalliance.org/2010/01/22/resilience-2011), and the third in Montpellier in May 2014 (http://www.resilience2014.org/).

Resilience has other antecedents, but my interest here is in showing the links between arid zone ecology in a very specific place (Australia), and the new global thinking. Holling's key paper provides a way to trace the local geographies of our planetary consciousness, and the way a scientific community develops its identity. This global identity is a key to the resilience idea. Holling is a Canadian, not an

Australian, and his work in temperate places in northern America has defined an important framework of modern population ecology, the non-linear dynamics of predator–prey systems. However, there is also a significant heritage for the concept of resilience in the boom-and-bust ecologies in the Australian desert and rangeland country. Even a global science can have heritage from one or several very particular places.

Because resilience is a concept that creates particular expertise useful in dealing with global change, it is useful to look briefly at the history of global change itself. Global change science is a suite of knowledge systems that include climate change science, global economics, demography, biodiversity science and global environmental change in all its facets. Resilience is closely allied with a range of global projects including the International Geosphere-Biosphere Program (IGBP) based in Stockholm; Analysis, Integration and Modeling of the Earth's System (AIMES) based in Boulder, Colorado, USA; and the International Human Dimensions Programme on Global Environmental Change (IHDP) based in Bonn, Germany. These groups have been influential in framing research agendas and findings for the Intergovernmental Panel on Climate Change (IPCC), the leading international body assessing the current state of knowledge of climate change and its potential environmental and socio-economic impacts. The IPCC was established under the auspices of the United Nations Environment Programme (UNEP) and is co-located with the World Meteorological Organization (WMO) in Geneva, Switzerland. The IPCC does not itself conduct research or monitor climate-related data or parameters, but it sponsors and directs research that informs its mission, including drawing on the work of other global change brokers including IGBP, AIMES, IHDP and the World Weather Research Programme, which has been part of WMO since 1980 (Robin and Steffen, 2007). A common language is important to discussions between different scientific groups and the policy advisors in IPCC.

Holling's resilience science in 1973

The prehistory of resilience and its role in global change science are complex. Here I focus on its trajectory from ecology, because this did much to theorise the space between science, policy and adaptive management. Holling's (1973) definition of resilience added a crucial new concept to ecological systems thinking, arguably the most important since the term *ecosystem* had been introduced (Tansley, 1935). Instead of looking at the stability of species and associations of species, resilience focused on measuring how ecosystems recovered or changed after shocks. No longer was science concerned to measure balanced ecosystems, idealised climax associations of biota or ecosystems in equilibrium. Rather, the new ecology focused on the surprises and the imbalances: it looked at transitions rather than at steady states.

Perhaps, too, the world was ready for this new idea in 1973. The year was significant for the international political context that was all about shocks, particularly the oil shocks that suggested that western society could not continue to grow indefinitely on the basis of fossil fuel economies. The Apollo missions of the late 1960s had raised consciousness of the singularity of the Earth with photographs from outer space looking back on the planet. The Club of Rome published *Limits to Growth* in 1972 and the United Nations conference on the environment was held in Stockholm the same year. The first oil shocks in 1973 underscored the limits to growth, and turned thinking toward new paradigms for the relations between humanity and nature. Scientific and social questions cast long shadows on each other, and created an atmosphere of support for an integrative imagination.

Holling was a mathematically-literate biologist asking broadly theoretical questions in different ways. He, like many others, was grappling with fitting biological world-views into the modelling paradigm of the IBP. Whereas the physical paradigm that underpinned the futures modelling was based on a universal world-view, ecological knowledge applicable to specific ecosystems and their feedback loops depended on the particularities of place. Holling (1973: 1) asked: Can we imagine a new world-view that 'concentrates not so much on presence or absence as upon the numbers of organisms and the degree of constancy of their numbers'? His predator–prey work had suggested considering not the survival of individuals but rather the survival of populations. Population biology and human ecology had trended in this direction since the 1920s. The idea of looking at species as whole societies was promoted by E. O. Wilson, for example, in his work on ants. Wilson focused on superorganisms, whole colonies, rather than the behaviour and survival of individuals (Wilson, 1971). In the same era, Richard Dawkins proposed the 'selfish gene' as the driver of evolutionary biology: he was also turning away from a focus on individual success, to the success of the gene pool as a whole (Dawkins, 1976).

Breaking the fascination with 'the balance of nature'

The idea of the balance of nature had driven the descriptions of natural history since antiquity. Balance has such intrinsic appeal that it often goes unquestioned. Linnaeus attempted a definition of it in the eighteenth century but it is more often implicit than explicit, as historian of biology Frank Egerton has explored (Egerton, 1973: 324). A. J. Nicholson and Carrington Williams had offered definitions of the balance of nature (in the twentieth century), but Egerton regarded these as *exceptions*, rather than the norm.

In ecology, Frederic Clements defined the stable or *climax* formation as balance in nature: climax is the last stage in ecological succession, 'when populations of all organisms are in balance with each other and with existing abiotic factors'. The 'climax formation' was the fully developed or adult version of an organic entity or community, a 'complex organism' (in Egerton, 1973: 344). Distinguishing between pure nature and nature tainted by human intervention made little sense in long-farmed environments like Arthur Tansley's Britain. The British ecologist developed a definition of an *ecosystem* to replace Clements's anthropomorphic notion of *community* (Tansley, 1935). Tansley's ecosystem replaced a theory of biological destiny determined by climate with a model of interrelatedness in nature that included biological and abiotic elements (including nutrient cycling and energy flows). The ecosystem became a key organising principle in ecology. It has robust explanatory power at local, regional and planetary scales – whether the subject is a pond, a prairie or Earth itself, so has proved much more versatile than Clements's 'vegetation community' (Worster, 1991: 278).

Nonetheless, ecological definitions of species abundance have continued to be described in terms of the dynamics of local communities, even though the boundaries of communities are not well-defined units. Particularly in animal ecology, the traditions established by Charles Elton persist. A community makes sense as a concept where local species compositions and population size are internally regulated within the niche. If, however, individual species in an assemblage respond individually to physical habitat factors (such as soil or altitude) – this is sometimes called Gleasonian dynamics – then the notion of an ecological community makes no sense at all (Sterelny, 2001: 438).

Holling grappled with a language to describe the fluctuations of populations. He did not want to describe them as settling into a balanced or climax ecosystem, because he wanted to take into account factors external to the immediate niche, including abiotic factors. Elton had famously asserted that 'the balance of nature does not exist, and perhaps never has existed' (Elton, [1927] 1966: 17). But despite Elton, Frederic Clements's climax theory continued to have strong appeal in plant ecology for much of the twentieth century although the strong climax theory had become muted, if not denied, by Clements himself by the 1940s (Egerton, 1973: 344; Cuddington, 2001; Kingsland, 2005).

Holling was keen to see if there were mathematically simple rules that did not need to assume the stability of an ecosystem. As he put it much later: simple but 'just sufficient'. He wanted a model that would allow ecosystems to adapt, to learn to deal with environmental change, including anthropogenic change. Since stability did not create the preconditions for adaptation, he excluded it from the model. 'High variability not low variability [is] necessary to maintain existence and learning' (Holling, 2006: 5–6; see Gunderson and Holling, 2002). This insight is critical to management, for it actually argues *against* the notion of corporate efficiency where duplication is treated as waste (Walker and Salt, 2006).

Arid zone science: post-war initiatives in ecology and management

Integrating knowledge across national boundaries was important to the politics of the post-war era, and was a priority of UNESCO (Robin and Steffen, 2007). One of the UNESCO initiatives was an Advisory Council on Arid Zone Research, an initiative of the newly independent India (UNESCO, 1953: 7). In Australia, C.S. Christian of the CSIRO was appointed Arid Zone Research Liaison Officer (in 1954) to co-ordinate the fragmented efforts of the states and Commonwealth in arid zone science. Although CSIRO Chairman, Ian Clunies Ross noted that:

'Australia has a disproportionately large share of the arid areas of the world . . . [with a] ratio of arid to non-arid land . . . [of] approximately three to one . . .' (1956: 2), answering the questions posed by the UNESCO Advisory Council proved challenging. Australia was proud of its internationally significant science, but its desert areas had been neglected. Science was difficult in a region that lacked roads and basic infrastructure. The UNESCO questions forced a different approach. From 1960 onwards, an increasing number of international arid zone experts visited the Australian deserts, often on Fulbright fellowships (Heathcote, 1987: 12). Their efforts were noted in the *Arid Zone Newsletter* (*AZN*) that Christian edited from 1956, which documented initiatives by CSIRO, state government agencies and universities, including international ones. *AZN* continued until 1987, and it was written to be read by people living in arid zone communities as well as policy-makers. Land systems surveys dominated the science until the 1960s, but after that there was an increasing interest in the physiology of desert-dwelling animals, particularly how they responded to stress under desert conditions. In 1961, two professors of zoology, Harry Waring from the University of Western Australia, and Jock Marshall, the Foundation Professor of Zoology at Monash University in Melbourne, led this research. Many of their graduates led the next generations of arid zone scientists in Australia and beyond. Research in Australian rangelands and deserts also began to contribute to degrees in overseas universities from the 1970s, including Utah, Tucson, Israel, Jodhpur and Guelph, each of which developed a particular specialty in arid zone science (*Arid Zone Newsletters*, 1973–1978; Robin, 2007: 114–122).

Deserts in Australia and Israel compared: Imanuel Noy-Meir

In 1973, in the same issue of the same journal as Holling's resilience paper, Argentinian-born Israeli ecologist, Imanuel Noy-Meir (1941–2009) published a seminal paper theorising a whole ecosystem approach to deserts (Noy-Meir, 1973). The *Annual Review of Ecology and Systematics* encouraged longer syntheses of major concepts and research programs. Resilience was the subject of its first paper that year, and desert ecosystems (part 1) was the second. Noy-Meir's paper defined *pulse and reserve* systems, a concept that theorised the effects of boom-and-bust climatic conditions on animals and plants in the highly variable Australian arid zone (Robin *et al.*, 2009). Like Holling's paper from 1973, Noy-Meir's is still cited today (see, for example, Morton *et al.*, 2011). The framework of the global IBP was evident in Noy-Meir's questions, just as it had been in Holling's. Noy-Meir's task was to study deserts for IBP, and the Australian desert was his key case. The pulse of desert life there depends on rain, but there can be years between events. When the rain finally comes, the reserve comes into play. Noy-Meir credited Australian rangeland scientist Mark Westoby with coining the term 'pulse and reserve system' (Noy-Meir, 1973: 30), and the uncertainty of the Australian system informed new theoretical approaches in these years and since (Noy-Meir, 1974, 1975; Morton *et al.*, 2011).

Pulse and reserve systems are not so much about the rain or boom conditions, but rather about the spaces between booms. Waiting out these resource-poor times between rains defines and limits desert life. Plants reserve their future in seeds. Animals like birds and kangaroos often move away in dry times – a nomadic strategy to avoid times of low resources. Other animals lie in wait: fish and frogs bury themselves and are awoken by the falling rain, when it finally comes. All have to be able to snap into active mode very quickly whenever the season arrives – not annually, but *whenever*. The best opportunists are those that jump quickest to the desert pulse. The capacity to switch between reserve mode and pulse mode is a survival advantage in deserts.

In 1973, Noy-Meir and Holling were both working as mathematical biologists. Both argued for a broader, rather than an increasingly specialised approach to their subjects. An interest in variability demanded looking for different drivers, modelling different questions. Holling had noted in his work on predation patterns that they fluctuated, but did not move in a particular direction, nor did they settle at a balance point. There was no equilibrium or climax in his data. So he turned his theoretical attention to 'the amplitude and frequency of oscillations'. His biologist's eye recognised that:

> An equilibrium centred view is essentially static and provides little insight into the transient behaviour of systems that are not near the equilibrium. Natural, undisturbed systems are likely to be continually in a transient state; they will be equally so under the influence of man.
>
> *(Holling, 1973: 2)*

Holling focused on changes external to the system, the surprises, and sought to map out the *persistence* of the relationships between elements of the ecosystems. By taking the ecosystem as a whole, his questions turned to its *sustainability* or, in some cases, total collapse. What Holling sought was mathematical precision, but with real-world ecological assumptions, to drive new models for ecologically complex situations.

A model based on the mathematics of standard physics 'generates neutral stability', he noted, 'but the [biological] assumptions are very unrealistic since very few components are included, there are no explicit lags or spatial elements, and thresholds, limits, and nonlinearities are missing' (ibid.: 5).

Rangeland management science: 1969–1989

There was a close relationship between the arid zone ecology of the 1950s in Australia and the professional rangeland management science that strengthened in the 1970s. Although the Australian desert or arid country was the geographical terrain of much of this new science, its intellectual forces were international. In 1973 and 1974, a United States/Australia Rangeland Panel held joint meetings in Tucson, Arizona, in 1973 and Alice Springs in 1974, under the auspices of the

United States Society for Range Management and the Australian Rangeland Society. Rangeland management was 'more than science': it was closely concerned with the economics and business directions of pastoral enterprises. Nonetheless, in practice, it was overwhelmingly biophysical in emphasis, and mathematical in methodology. In a major review of the *Rangeland Journal*, Witt and Page (2000) showed that 89 per cent of the nearly 300 articles published in the journal were biophysical in focus.

By the end of the 1980s, there was a growing awareness that management strategies might have *unexpected* consequences, and rangelands management scientists sought to theorise their causes. Irreversible change and the new systems created by shocks became the focus for their work, as they sought understandings of opportunism and drivers of non-equilibrium biophysical systems. Rather than managing for landscape stability, the research question of the late 1960s (see, for example, Ross, 1969, who was cited by Noy-Meir in 1973), Westoby, Walker and Noy-Meir observed that 'Vegetation changes in response to grazing have often been found to be not continuous, not reversible or not consistent' (1989: 268). Thus, a stable landscape may not be the product of a steady management strategy. Brian Walker, the third player in this paper, quickly became central to the resilience science story. Walker is a southern African rangeland scientist who moved to Australia in 1985 to head up CSIRO's Rangeland and Wildlife Division and is now an Australian citizen. Walker, like Holling, is a mathematically-literate, strongly applied biologist. Walker's focus for the past two decades has been Australian rangeland country – arid and semi-arid.

By the time Westoby, Walker and Noy-Meir were writing in 1989, theoretical ideas in ecology sought to model states and transitions in whole landscapes. Transitions could be triggered by natural events or management actions. They could occur quickly and may be irreversible. Most importantly, they observed that 'the system does not come to rest halfway through a transition' (ibid.: 268). Essentially the system flips, or moves into being a new system. The paper was the final nail in the coffin of climax ecosystems. The end of equilibrium theory pushed arid zone ecology, rangeland management science and resilience science closer together, and fostered a shared interest in the mathematics of heterogeneity.

Interdisciplinarity, networks and trustworthy places

Resilience thinking in the twenty-first century displays a high level of cross-publishing, co-authorship and cross-citation. The network is tight. The key names, Holling, Noy-Meir, Walker and Westoby recur, along with newer names like ecologist Carl Folke, later head of the Stockholm Resilience Centre, and global water expert Johan Rockström. Rockström was senior author of the planetary boundaries paper, mentioned above, which won him 'Swede of the Year' award in 2009 for his work on bridging science on climate change to policy and society (Rockström *et al.*, 2009). Two years later he was ranked the most influential person on environmental issues in Sweden.

The Resilience Alliance mimics a disciplinary scholarly community, but its aspirations are to go beyond disciplines, to integrate across the divides between society and ecology in thinking and practice. Since 1973, the word resilience has gained the imprimatur of technical integrative expertise, rather than being just a foundation concept of ecology. Its trajectory has been one of the forces that pushes the discipline of ecology towards policy relevance, along with ecological economics and concepts like 'ecological services' (Daily, 1997).

The seeds of policy relevance – of science for society, if you like – can be found in the imperial ecology of the Oxford School as far back as the 1920s and 1930s. Global-minded ecologists such as Julian Huxley and Charles Elton trained their students for work in the British dominions and colonies rather than to undertake ecological studies at home in England. Under Arthur Tansley, Oxford ecologists sought to theorise and systematise what had been natural history, in order to enlarge and improve the Empire. In historian Peder Anker's words, they wanted to use ecology to create 'a scientific management tool . . . to develop effective social systems . . . for society at large [and] for the administration of knowledge' (Anker, 2001: 78).

In 1921, Stephen Forbes urged the Ecological Society of America in his presidential address to consider humans (and their actions) as part of ecology as much as any other organism (Forbes, 1922). The discipline of human ecology was also highly implicated in Elton's Oxford School in the 1930s, though ecological science regarded itself as strictly biological, focusing on plants and animals other than humans, for much of the twentieth century. In a sense, human ecology returned to its original disciplinary fold in the 1990s with the foundation of the journal *Ecology and Society*, though there was great resistance within the ESA to this turn of events. The impetus for treating society and ecology together was furthered by the environmental crisis of the 1970s, and is most apparent in Michael Soulé's conservation biology, which in 1985 he defined as 'the science of crisis' (Soulé, 1985). The term *biodiversity* became famous through a Forum at the Smithsonian and the National Academy of Sciences in 1986 and its presence in academic and policy discourse has risen on a hockey-stick curve since then (Wilson, 1988; Farnham, 2007: 2).

Co-authorship and common institutions build trust. Ecologists need to know who is saying what, and precisely *where* it applies. The concept of resilience has developed global dimensions, including strong connections with the international policy community that has been central to the prominence of political concerns about climate change. Nonetheless, its origins in ecology are still evident. In their 2001 paper, Carpenter and his colleagues selected two major case studies to show how resilience thinking worked in an SES. The first was the Great Lakes regions on North America – a lacustrine system exploited for agriculture. The second resilience case study was the rangeland country of western New South Wales – semi-arid to arid country exploited for pastoralism. In the twenty-first century, there was more emphasis on economics and the lifestyles of the people, including their comparability as 'democracies with traditions of science-based management'.

The patchy and variable arid zone climate was part of the matrix. Variability is becoming something that is now increasing all over Earth with climate change, evidence is showing, and this gives desert science a new global mission (Carpenter *et al.*, 2001: 768). Like Holling three decades earlier, they describe two sorts of puzzle: those of 'self-contained ecosystems' (with linear effects) and those of open systems (with patchy effects), including those with human variables, which recover from shocks only because 'reinvasion' (repopulation from a fresh source) is possible (Holling, 1973: 10). Lakes and rangelands are the examples again in Carpenter *et al.*, just as they were in Holling.

As we have moved from *equilibrium* to *panarchy*, from biophysical to social ecological systems, we now see the natural world as cultural (Gunderson and Holling, 2002). There is fluidity between ecology and society. Such a world-view is changing what is meant by 'science', perhaps returning it closer to its origins in the Latin *scientia*, and certainly more in sympathy with the German concept of *Wissenschaft* that persisted right through the divides between science and humanities much discussed in Anglophone circles (Snow, 1959). Policy-making demands a knowledge base that is more than the sum of standard disciplines like economics and ecology. There is a new role for geography, history and all the humanities in understanding the global environment. As place shapes science as well as society, it is the peripheral geographies and exceptional places that drive the intellectual direction. Exceptions at the edge of survival like the Australian desert have pushed the mathematics of planetary survival and the modelling and imagination of the future through the concept of resilience.

References

Anker, P. 2001. *Imperial Ecology: Environmental Order in the British Empire, 1895–1945*, Cambridge, MA, Harvard University Press.

Bhatanacharoen, P., Greatbatch, D. and Clark, T. 2011. The Tipping Point of the 'Tipping Point' Metaphor: Agency and Process for Waves of Change, University of Durham Business School, unpublished paper.

Brand, F. S. and Jax, K. 2007. Focusing the Meaning(s) of Resilience: Resilience as a Descriptive Concept and a Boundary Object. *Ecology and Society* 12, Article 23. Available at: http://www.ecologyandsociety.org/vol12/iss1/art23/.

Braudel, F. 1995. *A History of Civilizations* (original French 1987; trans. R. Mayne 1993), New York, Penguin.

Broecker, W. S. 1987. Unpleasant Surprises in the Greenhouse? *Nature*, 328,123–126.

Carpenter, S., Walker, B., Anderies, J. M. and Abel, N. 2001. From Metaphor to Measurement: Resilience of What to What? *Ecosystems*, 4, 765–781.

Cittadino, E. 1980. Ecology and the Professionalization of Botany in America, 1890–1905. *Studies in the History of Biology*, 4, 171–198.

Clunies Ross, I. 1956. Introduction. *Arid Zone Newsletter (AZN)*, 1, 1–2.

Crosby, A. 1986. *Ecological Imperialism: The Biological Expansion of Europe, 900–1900*, Cambridge, Cambridge University Press.

Crutzen, P. J. and Stoermer, E. F. 2000. The 'Anthropocene'. *IGBP Newsletter*, 41, 17–18.

Cuddington, K. 2001. The 'Balance of Nature' Metaphor and Equilibrium in Population Ecology. *Biology and Philosophy*, 16, 463–479.

Daily, G. C. 1997. *Nature's Services: Societal Dependence on Natural Ecosystems*, Washington, DC, Island Press.

Dawkins, R. 1976. *The Selfish Gene*, New York, Oxford University Press.

Deveson, A. 2003. *Resilience*, Crows Nest, NSW, Allen & Unwin.

Egerton, F. N. 1973. Changing Concepts of the Balance of Nature. *The Quarterly Review of Biology*, 48, 322–350.

Elton, C. 1958. *The Ecology of Invasions by Animals and Plants*, London, Methuen & Co.

Elton, C. [1927] 1966. *Animal Ecology*, Seattle, University of Washington Press.

Farnham, T. J. 2007. *Saving Nature's Legacy: Origins of the Idea of Biological Diversity*, New Haven, CT, Yale University Press.

Folke, C. 2006. Resilience: The Emergence of a Perspective for Social-Ecological System Analyses. *Global Environmental Change*, 16, 253–267.

Forbes, S. A. 1922. The Humanizing of Ecology. *Ecology*, 3, 89–92.

Forbes, S. A. [1887] 1925. The Lake as a Microcosm. *Bulletin of the Peoria Scientific Association*, 1887: 77–87; reprinted in *Bulletin of the Illinois State Natural History Survey*, 15, 537–550.

Golley, F. B. 1993. *A History of the Ecosystem Concept in Ecology: More Than the Sum of the Parts*, New Haven, CT, Yale University Press.

Goodall, H. and Cadzow, A. 2009. *Rivers and Resilience: Aboriginal People on Sydney's Georges River*, Sydney, UNSW Press.

Gunderson, L. H. and Holling, C. S. (eds) 2002. *Panarchy: Understanding Transformations in Human and Natural Systems*, Washington, DC, Island Press.

Hastrup, K. 2009. Destinies and Decisions: Taking the Life-World Seriously in Environmental History. In: Sörlin, S. and Warde, P. (eds) *Nature's End*, London, Palgrave Macmillan, pp. 325–342.

Heathcote, R. L. 1987. Images of a Desert. *Australian Geographical Studies*, 25, 3–25, 12.

Holling, C. S. 1973. Resilience and Stability of Ecological Systems. *Annual Review of Ecology and Systematics*, 4, 1–23.

Holling, C. S. 2006. A Journey of Discovery. Available at: http://www.resalliance.org/index.php/holling_memoir.

Huenneke, L., Glick, D., Waweru, F. W., Brownell, R. L. and Goodland, J. R. 1988. Scope Program on Biological Invasions: A Status Report. *Conservation Biology*, 2(1), 8–10.

Hulme, M. 2013. *Exploring Climate Change through Science and in Society: An Anthology of Mike Hulme's Essays, Interviews and Speeches*, London, Routledge Earthscan.

Kingsland, S. E. 2005. *The Evolution of American Ecology, 1890–2000*, Baltimore, MD, Johns Hopkins University Press.

Leakey, R. and Lewin, R. 1996. *The Sixth Extinction: Patterns of Life and the Future of Humankind*, New York, Anchor Books.

Lindenmayer, D. and Fischer, J. 2007. Tackling the Habitat Fragmentation Panchreston. *Trends in Ecology and Evolution*, 22(3), 127–132.

Lindsay, R. W. and Zhang, J. 2005. The Thinning of Arctic Sea Ice, 1988–2003: Have We Passed a Tipping Point? *Journal of Climate*, 18, 4879–4894.

Livingstone, D. N. 2003. *Putting Science in its Place: Geographies of Scientific Knowledge*, Chicago, University of Chicago Press.

Luthar, S.S. 2006. Resilience in Development: A Synthesis of Research Across Five Decades. In: Cicchetti, D. and Cohen, D. J. (eds) *Developmental Psychopathology*, Vol. 3, *Risk, Disorder and Adaptation* 2nd edn, Hoboken, NJ, John Wiley & Sons Inc, pp. 739–795.

Meadows, D. H., Meadows, D. L. and Randers, J. 1972. *The Limits to Growth: A Report for the Club of Rome Project on the Predicament of Mankind*, London, Club of Rome.

Morton, S. R., Stafford Smith, D. M., Dickman, C. R., Dunkerley, D. L., Friedel, M. H., Mcallister, R. R. J., *et al.* 2011. A Fresh Framework for the Ecology of Arid Australia. *Journal of Arid Environments*, 75, 313–329.

Noy-Meir, I. 1973. Desert Ecosystems: Environment and Producers. *Annual Review of Ecology and Systematics*, 4, 25–51.

Noy-Meir, I. 1974. Desert Ecosystems: Higher Trophic Levels. *Annual Review of Ecology and Systematics*, 5, 195–214.

Noy-Meir, I. 1975. Stability of Grazing Systems: An Application of Predator–Prey Graphs. *Journal of Ecology*, 63, 459–481.

Olsson, P., Folke, C. and Birkes, F. 2004. Adaptive Co-management for Building Resilience in Social–Ecological Systems. *Environmental Management*, 34, 75–90.

Pickett, S. T. A., Cadenasso, M. L. and Grove, J. M. 2004. Resilient Cities: Meaning, Models, and Metaphor for Integrating the Ecological, Socio-Economic, and Planning Realms. *Landscape and Urban Planning*, 69, 369–384.

Porter, T. M. 1995. *Trust in Numbers: The Pursuit of Objectivity in Science and Public Life*, Princeton, NJ, Princeton University Press.

Richardson, D. M. 2011. *Fifty Years of Invasion Biology: The Legacy of Charles Elton*, New York and London, Wiley-Blackwell.

Robin, L. 2007. *How a Continent Created a Nation*, Sydney, UNSW Press.

Robin, L. 2013. Histories for Changing Times: Entering the Anthropocene? *Australian Historical Studies*.

Robin, L., Heinsohn, R. and Joseph, L. (eds) 2009. *Boom and Bust: Bird Stories for a Dry Country*, Melbourne, CSIRO Publishing.

Robin, L. and Steffen, W. 2007. History for the Anthropocene. *History Compass*, 5(5), 1694–1719.

Rockström, J., Constanza, R., Svedin, U., Falkenmark, M., Karlberg, L., Walker, B., *et al.* 2009. A Safe Operating Space for Humanity. *Nature*, 461, 472–475.

Ross, M. 1969. An Integrated Approach to the Ecology of Arid Australia, *Proceedings of the Ecological Society* 4, 67–81.

Schneider, D. W. 2000. Local Knowledge, Environmental Politics, and the Founding of Ecology in the United States: Stephen Forbes and 'The Lake as a Microcosm'. *Isis*, 91, 681–705.

Shapin, S. 1998. Placing the View from Nowhere: Historical and Sociological Problems in the Location of Science. *Transactions of the Institute of British Geographers*, New Series, 23, 5–12.

Snow, C. P. 1959. *The Two Cultures and the Scientific Revolution*, The Rede Lecture (7 May 1959), Cambridge, Cambridge University Press.

Soulé, M. E. 1985. What Is Conservation Biology? *Bioscience*, 35, 727–734.

Steffen, W., Grinevald, J., Crutzen, P. and McNeill, J. 2011. The Anthropocene: Conceptual and Historical Perspectives. *Philosophical Transactions of the Royal Society A*, 369, 842–867.

Sterelny, K. 2001. The Reality of Ecological Assemblages: A Palaeo-ecological Puzzle. *Biology and Philosophy*, 16, 437–461.

Tansley, A. G. 1935. The Use and Abuse of Vegetational Terms and Concepts. *Ecology*, 16, 284–307.

UNESCO, 1953. *Reviews of Research on Arid Zone Hydrology (I)*, Paris, UNESCO.

Walker, B., Holling, C. S., Carpenter, S. R, and Kinzig, A. 2004. Resilience, Adaptability and Transformability in Social–Ecological Systems. *Ecology and Society*, 9, 5. Available at: http://www.ecologyandsociety.org/vol9/iss2/art5.

Walker, B. and Salt, D. 2006. *Resilience Thinking: Sustaining Ecosystems and People in a Changing World*, Washington, DC, Island Press.

Westoby, M., Walker, B. and Noy-Meir, I. 1989. Opportunistic Management for Rangelands Not at Equilibrium. *Journal of Range Management*, 42, 266–274.

Wilson, E. O. 1971.*The Insect Societies*, Cambridge, MA, Belknap Press of Harvard University Press.

Wilson, E. O. (ed.) 1988. *BioDiversity*, Washington, DC, National Academies Press.

Witt, B. and Page, M. 2000. Is Rangeland Research Driven by Discipline?: An Analysis of the *Rangeland Journal* 1976–1999. In: Nicolson, S. and Noble, J. (eds) *Proceedings of the Centenary Symposium*, Broken Hill, NSW, Australian Rangeland Society (21–24 August), 45–48.

Worster, D. 1991. *Nature's Economy: The Roots of Ecology*, rev. 2nd edn, Cambridge, Cambridge University Press.

Zalasiewicz, J., Williams, M., Haywood A. and Ellis, M. 2011. The Anthropocene: A New Era of Geological Time? *Philosophical Transactions of the Royal Society A*, 369, 835–841.

5

LANDSCAPES OF THE ANTHROPOCENE

From dominion to dependence?

Eric Pawson and Andreas Aagaard Christensen

In his now–classic book, *The Ecology of Invasions by Animals and Plants*, Charles Elton writes of returning home to Britain from Wisconsin shortly before the Second World War. He had with him some large American acorns which he intended to keep as mementoes. A few days later, chafer beetle grubs began to emerge unexpectedly from the acorns, forcing him to drop them all promptly into boiling water. The telling observation is that which follows: 'When the customs officer had asked me if I had anything to declare, it never occurred to me to say "acorns", and I am not sure if he would have been interested if I had' (Elton, 1958: 111).

This anecdote may seem remarkable today when border biosecurity controls are increasingly strict. But the free mobility of plant material in the hands of humans was for long a characteristic feature of the Anthropocene, being the basis of agricultural experimentation and improvement throughout much of the world. These were intimately connected to practices of 'enclosure', which saw large parts of northern Europe being reordered spatially so as to align rural land use with the emerging capitalist social order (Clout, 1998). A parallel development took place in the European colonies at the time, with the export of the modern capitalist landscape model across the globe, including to the Americas and Australasia. Elton himself wrote of New Zealand that 'No place in the world has received for such a long time such a steady stream of aggressive invaders' (Elton, 1958: 89), due to the permeability of its territorial borders for many years. This enabled the assembly of what Alfred Crosby (1986: 270) described as a 'portmanteau biota' of plant and animal species as the basis of making newly improved landscapes.

The aim of this chapter is to explore the dramatic increase in the power of human agency over the environment through an analysis of landscape change. It discusses the processes that have shaped new landscapes in the capitalist world before focusing on one place that is characteristic of the shifting balance of ecological agency in favour of humans during the Anthropocene. Banks Peninsula on the east coast of New Zealand's South Island was first settled by Polynesian peoples within the last few hundred years. The nature of their footprint contrasts with the dramatic change wrought by Europeans since the 1840s, when indigenous forests were transformed into improved landscapes of sown grass. The chapter is shaped by a broad question: What can be learned from this place about the ways in which people have exercised and are coming to terms with what Gibson-Graham and Roelvink describe as our 'gargantuan agency' and 'almost unbearable level of responsibility' in the Anthropocene (2009: 321)? It concludes with a discussion of the concept of 'middle landscapes' as one means by which the planetary dominion of humanity might be tempered with a realization of its dependence on terrestrial ecosystems for continued survival.

Landscapes of the Anthropocene

Since the term 'Anthropocene' was first proposed by the Dutch atmospheric chemist Paul Crutzen in 2000, it has been widely adopted as a dramatic, informal metaphor of the effect of human agency in global environmental change dynamics. In 2008, members of the Stratigraphy Commission of the Geological Society of London asked: 'Are we now living in the Anthropocene?', and concluded on the evidence of changes to physical sedimentation, carbon cycle perturbation and temperatures, biotic extinctions and migrations and surface ocean acidity, that there was sufficient grounds to recognize it 'as a new geological epoch' (Zalasiewicz *et al.*, 2008: 8). This potential recategorization was discussed at a one-day meeting of the Geological Society in 2011. More formally, it was placed before a working group of the International Commission on Stratigraphy, which is due to report by 2016.

The issues that concern this group include when and with what criteria the boundary should be drawn with the Holocene, that period of post-glacial geological time that ushered in conditions suited to the development of human agriculture and rapid population expansion. Most commentators agree with Crutzen's original use of a social as well as scientific marker, being the onset of the European Industrial Revolution from the late eighteenth century. Significantly, this begins to give content to the 'Anthropocene' in specific and living ways, taking it well beyond the geological into other disciplinary realms. The 2008 study cited human landscape transformation as the cause of an observable increase over the last 200 years in the denudation rate, above that of 'natural sediment production by an order of magnitude' (ibid.: 8). The nature of this landscape transformation has attracted the attention of historically minded scholars for just as long, but crucially this has been framed by how the concept of landscape itself is understood.

Landscape is an ambiguous word, loaded with a range of meanings, but with the consistent function of acting as a lens through which the relations between human cultures and nature are scrutinized. As a geographical term, Wylie (2011) proposes that work on landscape can be identified within three broad traditions: landscape (1) as material record; (2) as a way of seeing; and (3) as dwelling. The first is most closely associated with the American cultural geographer Carl Sauer (1963), who sought to describe the tangible morphology of landscapes as a product of the active role of people in shaping the world. The second has been built from the more recent work of British cultural geographers, focusing on the cultural politics of landscape. It is a body of work that aims to understand how landscapes are shaped in ways that privilege 'distanced, objective and penetrating' ways of seeing (Pickles, 2004: 83). Third, the diverse approaches to landscape as dwelling seek to reclaim both its materiality, as well as to pay attention to the ongoing association of labour and land – or the work of landscaping – that is discounted by the Sauerian tradition of treating landscape as 'a tangible externality' (Wylie, 2011: 303). Hence Ingold (2000: 191) argues that landscape is neither 'a picture in the imagination' nor 'an alien and formless substrate awaiting the imposition of human order'. Rather it signifies the temporary and dynamic result of a dialectical relationship between human agency and the lived-in world.

Kenneth Cumberland and Andrew Hill Clark, both with close links to Sauer, provided the first academic expositions of New Zealand landscape change in the colonial period that fostered the Anthropocene conditions of today. Cumberland highlighted the condensed time scale applying in New Zealand in an article arguing that 'What in Europe took twenty centuries, and in North America four, has been accomplished in New Zealand within a single century – in little more than one full lifetime' (1941: 529). Clark published a well-known study of the South Island, the central concern of which Elton was subsequently to echo, entitled *The Invasion of New Zealand by People, Plants and Animals* (1949). More recently, writing on the theme of landscapes of empire, W. J. T. Mitchell has suggested that 'Landscape might be seen . . . as something like the "dream-work of imperialism"' disclosing 'both utopian fantasies of the perfected imperial prospect and fractured images of unresolved ambivalence and unsuppressed resistance' (Mitchell, 1994: 10). Materially, this 'dream-work' was mobilized by three practices of landscape production: enclosure, improvement and biotic exchange. They are each deeply implicated in the historical geography of industrial capitalism through what David Harvey has called 'accumulation by dispossession' (Harvey, 2003).

Classical liberal thinking fashioned a basic restructuring of social relations with nature. It was John Locke who laid the ideological and discursive foundations for what is in effect the political philosophy of the capitalist social order in the Anthropocene. For Locke, a just and efficient order was one based on individual property rights over land, guaranteed by the state, as opposed to traditional conservatism, hereditary privilege and absolutist government. He ascribed value in nature as being conferred through the application of labour, and conversely denigrated 'unimproved' nature as valueless, or lying 'waste'. His was both an instrumental approach to nature and a characterization of society as based on the control of land by those individuals who both own and arrange for it to be worked (Barry, 1999; McCarthy and Prudham, 2004). To attain this, it was necessary to enclose resources held in common, including those for grazing, hunting and fishing. This freed nature from complex customary restraints, at the same time as enabling its profit-making promise (Goldstein, 2013). In Europe, not all common rights were extinguished. In contrast to colonial settings, which were strictly privatized, rights of way across landscape were often maintained.

The enclosures of British 'wastes' and those of the colonial periphery have generally been the subject of discrete literatures. For Karl Marx, they were hardly separate, since enclosure – and its proxies of commodification, privatization and dispossession – were fundamental to the emergence of capitalist logic and its geographical expansion. Over-accumulation, or the lack of opportunities for profitable investment, could be resolved through the opening up of demand for investment and consumer goods via imperialist activity (Harvey, 2003). Meanwhile, to Edward Gibbon Wakefield, the theorist of colonial promotion, it was not over-accumulation so much as over-population that required the safety valve of geographical expansion (Fairburn, 2012). Marx saw Wakefield's contribution as aimed 'at manufacturing wage-labourers in the colonies' through exploitation and injustice resulting from unequal class relationships (Marx, 1976: 932); Wakefield and his contemporary liberal colleagues represented it as efficiency and order. Wakefield's mechanism of the 'sufficient price' of land (which was to provide funds to support the social reproduction of new societies as well as a fluid supply of wage labour to improve property for its new owners) depended upon unimproved land being deemed 'waste'. That is, it was to be acquired from its indigenous owners for next to nothing and on-sold at a much higher price to new colonizing occupants. In the liberal discourse of both colonies and 'home', improvement was therefore the route to the realization of nature's potential.

A system of land titles and tenure provided the spatial and social building blocks of this new settler economy, being the basis upon which improvement and enclosure could be implemented. Enclosure was enforced and practised using the technologies of surveying and fencing, which allowed for effective management and containment of flows of living bodies, people and animals alike (Christensen, 2013; Goldstein, 2013). They delineated private property rights in territory at the same time as excluding those dispossessed. Fences also enabled the parcelling of land for removal of its indigenous land cover, and its replacement by introduced plant crops, the most significant of which in temperate latitudes has been sown grass (Brooking and Pawson, 2011). Biotic exchange, both in place and across space, has therefore been a key to improvement undertaken to generate surplus value during the Anthropocene. Ideologically it was underpinned, in New Zealand at least, by a specific reading of Darwinian evolutionary theory that appeared both to explain and to legitimate the speed with which indigenous vegetation was replaced by a more vigorous northern hemisphere biota (Livingstone, 2005). The residue of such beliefs has long persisted in the management of New Zealand landscapes.

Banks Peninsula: a landscape of 'improvement'

Banks Peninsula is an area of about 1,000 square kilometres; geologically it is the product of late Tertiary volcanic activity. It was originally named 'Banks's Island'

by Captain Cook, after the botanist aboard the *Endeavour*. To Cook, the then
tenuous connection of the Peninsula to what were later to be the 'Canterbury
Plains' was not apparent from offshore. Today it is one of the more richly textured
landscapes of the South Island, a finely grained mosaic of small settlements, farms,
sheep pastures, exotic trees and regenerating bush worked into the steep slopes of
the worn-down calderas of Akaroa and Lyttelton harbours (Figure 5.1). It is a
landscape that is largely silent about the conditions of its own production, but one
that well illustrates the result of accumulation by dispossession and the attendant
practices of enclosure, improvement and biotic exchange.

The pattern of land acquisition by Europeans from Ngāi Tahu, the *iwi* (tribe)
who hold *mana whenua* (tribal authority) over much of the South Island, is complex.
It reflects the meeting of two different political orders, uses of space and tenure
systems (Stevens, 2013). The Māori relation between people and land was based
on use value expressed in customary, multifunctional rights that often overlapped
both socially and territorially. It was a system presented in narrative form rather
than one represented through an object or text, such as a map or land title. Rights
did not reside with proprietary landholders, but were invested more widely through
tribal structures (Christensen, 2013). There was limited appreciation of these
differences by representatives of the British Crown in New Zealand, who were
instructed to obtain signatures in the South Island to the Treaty of Waitangi in
1840. The Crown land purchases that followed, codified in sketchy maps, treated
Māori territory as *tabula rasa*, while Māori understood the rich detail of oral

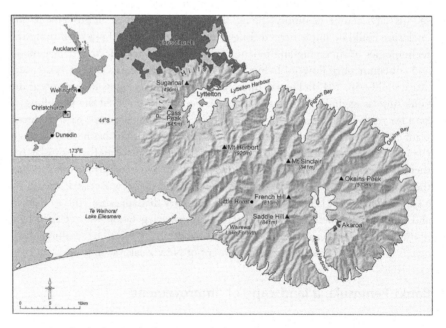

FIGURE 5.1 Banks Peninsula: location and place names.

negotiations, encompassing that which they wished to keep, to be the authoritative contract. Cutting across this was the involvement of French colonizers, whose interest had been piqued through Peninsula whaling operations in the late 1830s. Underlining the epistemic and ontological distance between the parties, an important link in the French claim was a blank sheet of paper signed by some Ngāi Tahu, upon which the French representative inscribed a written contract 'some days later' (Waitangi Tribunal, 1991: 532).

An assessment of this process of dispossession was undertaken by the Waitangi Tribunal in the 1990s in its statutory role of investigating Māori grievances arising from alleged Crown breaches of the Treaty. One of the statements of grievance in that part of the Ngāi Tahu claim relating to Banks Peninsula was:

> That the Crown has failed to ensure the adequate protection of the natural resources of Banks Peninsula; that it has allowed the wholesale destruction of the forests and other natural vegetation to the detriment of native fauna, water quality and soil conservation, and that the resulting siltation of stream beds and tidal waters has been to the detriment of fish and birdlife; that the Crown has allowed excessive pollution of Wairewa (Lake Forsyth) so that this great inland fishery and eel resource is now almost extinguished; and that it has allowed the depletion of kaimoana in the bays, harbours and coasts through pollution and excessive exploitation.
>
> *(Waitangi Tribunal, 1991: 527)*

Here voice is given to the 'unsuppressed resistance' to the remaking of landscape as imperialism's dream-work.

Several attributes of this statement of claim are worth comment. It is a relational description of landscape, in which the loss of customary resources such as seafood (*kaimoana*), eels and birds is linked to the 'wholesale destruction of forests' and 'the resulting siltation of stream beds and tidal waters'. The conditions necessary for the maintenance of these resources would, in pre-European times, have been regulated in specific ways through Māori social structures. This was critical as Māori subsisted in a land without mammals. Regulation developed fluidly over time in response to ongoing human impacts, both intentional and accidental. A quarter or more of the forest cover had been destroyed before European arrival (Figure 5.2), and with that about 30 bird species out of 100 vanished (Wilson, 2008). It is likely that Māori used fire to clear land for horticulture, and to open up clearings to facilitate the hunting of flightless birds.

The Tribunal statement is therefore a claim about the remembrance of Māori environmental dependence. It also adopts a particular vocabulary, a notable word being 'forests'. The New Zealand forests were often tall, tangled and dense compared to the familiar woodlands of Europe. They represented both a discursive and a practical challenge to the new settlers. A common response to the first was to adopt a specific term, 'bush', which conveyed the sense of difference (Johnston, 1981). The second was important too, as trees were something other than living

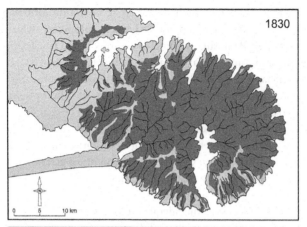

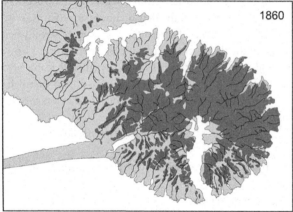

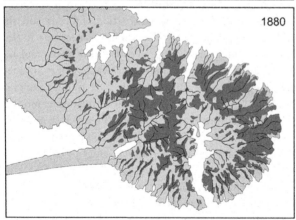

FIGURE 5.2 Clearing the bush: estimates of its extent in 1830, 1860 and 1880. By the 1920s, less than 1 per cent of the Peninsula still had old-growth forest cover. The maps are broad estimates based on contemporary surveys.

Source: Wilson (2008, 256–257).

forms to support birds, insects and other plants. In the 'wooden world' of the colonial project, for which buildings, fences and transport infrastructure (including bridges, railway sleepers and telegraph poles) were essential, they were a commodity (Wynn, 2013).

The bush of Banks Peninsula was accessible to the Canterbury Association's main town of Christchurch and its port of Lyttelton, which were surveyed in 1850. An assessment at that time was that there was sufficient bush to last 20 Canterbury colonies for centuries, such was its apparent impenetrability (Petrie, 1963). In fact most was cut out and burned in the half century to 1900, by sawyers and sawmillers working from the coastline up the valleys and then into the steep slopes up towards the ridges (Figure 5.2). Timber was cut on freehold sections to which title had been granted by the Crown, whereas timber licences were required for cutting on Crown land. It was the large timber trees that were most sought, such as the rimu of Little River bush, and totara and matai that were to be found across the Peninsula. These were valued for building purposes, and would be cut, and then rolled down slopes or hauled on wooden rails to a saw pit, and later to steam-powered mills. The cut over-bush was usually burned at the end of the summer, when it was most likely that a clean, 'white' burn would result. The fires were impossible to confine and much uncut timber was lost this way: in the 1880s, the Little River district was described as 'the valley of a thousand fires' (in Petrie 1963: 80).

The Māori population of the Peninsula was effectively dispossessed through this process: the French deed, subsequently subsumed by the Crown, allowed them only a small number of reserves, amounting to a few hundred hectares in all. Their original desire to gain the benefits of trade by allowing Europeans to settle in close proximity was not realized, here as elsewhere (Belich, 1996). There was, however, some concern from a minority of settlers about bush clearance. On 7 October 1868, Thomas Potts, the Member of the House of Representatives for Mount Herbert (the Peninsula constituency in the Wellington lower chamber of Parliament), moved that government 'ascertain the present condition of the forests of the Colony, with a view to their better conservation'. He said in an eloquent speech that he had 'often seen Banks' Peninsula covered, for weeks together, with thick and lurid smoke' (Potts, 1868: 188–189).

Potts was a local naturalist, but his concern was wider than this. He had read George Perkins Marsh's book *Man and Nature*, published in New York in 1864, and had drawn from it not only an understanding of what he called the 'barbarous improvidence' of wasted timber resources but also the 'mischievous results' in terms of 'excessive inundations' (Potts, 1868: 188). His intervention had some impact: Marsh's book was discussed by the New Zealand Institute and a few years later Julius Vogel, then Premier, aware of Potts' contribution, undertook a tour of the South Island's forests. He then successfully introduced a Forests Bill, an attempt at improved management of the national timber resource (Wynn, 1979).

The picture of landscape that met with greatest favour by the rapidly growing European population was, however, strictly productivist. This was keenly expressed by another Peninsula resident, the journalist H. C. Jacobson. He wrote:

> Where the mighty totaras once proved a home for thousands of native birds, good succulent grasses nourish stock which brings wealth to their proprietors and revenue to the Colonial Government. Many regret the passing away of the old order of things . . . but we cannot help fancying that to the thinking person the present landscape is far more gratifying. True gloomy Rembrandt-like shadows have disappeared, and . . . in the stead of the past beauties are smiling slopes of grass . . .
>
> *(Jacobson, 1893: 229)*

In about 1925, when the transition to grass was as complete as it was ever to be, a local photographer recorded a panoramic image of the smiling slopes above Akaroa harbour (Figure 5.3). This was nature improved.

A landscape of grass

The significance of Jesse Buckland's photograph in Figure 5.3 is that it represents in microcosm one of the great landscape transformations of the Anthropocene. This is the transition from trees to grass that was engineered in many parts of newly settled European world (Wood and Pawson, 2008). It was driven by a move towards intensified and extensified animal-based husbandries, as the growing urban populations of northern Europe and America began to consume vastly more protein in the form of meat and cheese. The grass in the photograph is not indigenous, but is likely to be an introduced exotic species, cocksfoot, known in the United States as orchard grass. New Zealand farmers sometimes obtained grass seed by mail from English seed companies, but more usually from local companies or nurseries that in turn sourced it from local surplus or overseas. Locally sourced

FIGURE 5.3 The improved slopes of Akaroa harbour, denuded of bush (compare to Figure 5.2).

Source: Photograph by Jesse Buckland, c. 1925, from Okains Peak, looking south. She appears to have carefully composed this scene, with three standing tree skeletons in the foreground, along with other burnt detritus that speaks of the removed forest.

Source: AK: 2003.18.2.27, Akaroa Museum.

seed was frequently contaminated with a range of accompanying weeds such as thistles. On the Peninsula, it was not sown, but rather broadcast by hand directly into the ashes of the bush fires. It would germinate in the autumn rains and by the next season a reasonable coverage could be expected (Petrie, 1963).

With grass came animals, and the ability to export surplus food and fibre in exchange for investment and consumer goods. Grass species were therefore a foundational component in international flows of biota (Brooking and Pawson, 2011). The nature of these flows has received some attention in recent years. In Crosby's (1986) view of an imperial world focused on a metropolitan centre, they are explicitly part of the armoury, with European people and their pathogens, of 'ecological imperialism'. The species carried were entrained in the work of naturalizing a particular type of order that in its clarity and on-the-ground delineation represents a classical liberal landscape patterned through proprietorial rights. More recently, it has been argued that a globalized world is more realistically understood as constituted from countless local places, in which case distinctions between centre and periphery blur (Lester, 2006; Pawson, 2008). A view of landscapes as relational in this sense reveals more complex networks of flows and more subtle ways in which processes of accumulation and landscape making have occurred. From this perspective, Banks Peninsula assumes a quite different position in what has been called the 'webs of empire' (Ballantyne, 2012).

Cocksfoot, *Dactylis glomerata*, was suited to the Peninsula. It grows well in shady places, like forest margins, but also tolerates dry conditions, such as occur locally in mid to late summer on what are free-draining soils. The earliest known paddocks of cocksfoot are recorded in Pigeon Bay in the mid-1850s. It rapidly became a popular crop in its own right, supplementing other forms of farm income, as the seed found a ready market in bush burn districts around New Zealand. Peninsula cocksfoot thereby provided the basis of the forest-to-grass transition in places like Taranaki and Manawatu in the North Island and in the Catlins in the southeastern part of the South Island. The trade represented more than the disposal of surplus. Cocksfoot has a high degree of intra-species variability and the local strain – which became known as 'Akaroa cocksfoot' – evolved relatively quickly. It was longer lived and also provided better all-year-round growth than overseas strains. It therefore became a distinctive commodity in its own right (Wood, 2008).

By the 1880s a substantial amount was being traded internationally (Figure 5.4). Although there was a lot of annual variation, depending on the amount of seed available, exports built up rapidly with the peak years being just before and after the First World War. The bulk of the seed was destined for Britain and Australia, but increasingly America became a good customer. The 'other' category in the graph includes Germany and Ireland. The significance of cocksfoot can only be understood in terms of how it was used in the repair of old pastures and the extension of new ones. Nineteenth-century grass seed mixes could be quite complex but by the 1880s had often converged around three plants: cocksfoot, perennial ryegrass, and white clover. In this context, cocksfoot played a valued role, its attributes balancing those of long-favoured ryegrass. It is a clumpy plant,

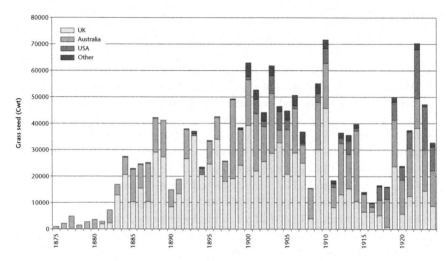

FIGURE 5.4 Exports of grass seed from New Zealand, 1875–1924. This is a good proxy for export of Banks Peninsula cocksfoot as only Lyttelton among New Zealand ports exported significant quantities of grass seed.

Source: Brooking and Pawson (2011: 128).

between which the ryegrass could grow, and provides herbage earlier in the season. Cocksfoot is also persistent whereas, in the 1880s, the so-called 'ryegrass controversy' in Britain raised doubts about whether ryegrass had this quality (Wood, 2008).

Cocksfoot was harvested by gangs of men, women, children and animals, who assembled each season in late summer on the wharves of Lyttelton. Figure 5.5, another Jesse Buckland photograph, shows a gang at work in the mid-1920s. A portable threshing machine has been set up on a flat site below a bush remnant; winnowing is taking place alongside. The grassy slopes, from which the seed heads have been cut by hand, stretch into the distance at the right. A woman fills and sews sacks of seed, which are then loaded onto horses. A dog rests on sacks yet to be collected. This is a scene posed for the photographer but its shows the sequence of material tasks required to process cocksfoot in the fields. These are tasks of landscaping, what Kenneth Olwig refers to as 'performing upon landscape', or undertaking the work of landscape maintenance. Landscape in this sense is 'a woven material created through the merging of body and senses that occurs in dwelling' (Olwig, 2008: 83–84).

The photograph reveals little about the temporality of landscape (Ingold, 2000). Paddocks intended for seed production were shut up between early spring and late summer to enable the seed to mature, then re-opened for cows in the autumn, when the grass began to produce new and palatable shoots. The annual harvest took place between January and March, depending on the aspect and elevation of the valley slopes. Seed spillage enabled the crop to extend between the clumps and maintain the pasture. By the turn of the twentieth century, the Peninsula was known as 'cow and cocksfoot' country (Petrie, 1963), and most valleys had a small cheese or butter factory. Cheese and butter production were strictly regulated by the state, but grass seed production was not. It was, however, subject to increasing commercial oversight as seed companies in Christchurch and the southern city of Dunedin sought to distinguish their product in a crowded market place. They did this by means of various metrological claims concerning relative freedom from contamination and germination rate (Brooking and Pawson, 2011).

The Peninsula industry began to shrink in the mid-1920s. Akaroa cocksfoot was expensive as the topography meant that production would always be reliant on a high degree of hand processing. Increasingly the industry began to shift to new seed-producing areas on the much more easily worked Canterbury Plains. The local strain was also under challenge from new ones developed overseas, notably in Denmark. In 1926, the highly regarded British grassland scientist George Stapledon visited the Peninsula while on leave from the Welsh Plant Breeding Station in Aberystwyth, Wales. Stapledon had established the station in 1919 in response to government concerns about the state of British pastures. He subsequently developed the S.37 and S.143 strains from Akaroa seed (Brooking and Pawson, 2011). By then, cocksfoot was losing its popularity in grass seed mixes as new ryegrass strains had finally resolved the lingering controversy of the 1880s. Nonetheless, this episode reflects another characteristic of Anthropocene land-scapes, the global exchange of genetic material through complex and contingent webs in order to develop plants best adapted to conditions prevailing in any particular site of production.

Transitioning to a middle landscape?

In the 1920s, less than 1 per cent of the Peninsula's land area remained under old growth forest (Wilson, 2008). The frontier of bush clearance moved elsewhere, as pastoral farming developed in the hill country of the North Island in response to

FIGURE 5.5 A cocksfooting gang at work. Photograph by Jesse Buckland, c. 1925.

Source: AK: 2003.18.2.30, Akaroa Museum.

the growth of the British market for frozen lamb. The production of these 'improved' landscapes, often newly prone to surface erosion, did not impress all thinking people (Cumberland, 1941). A well-known essay published in 1952 portrayed the land as being used 'not for farming but mining', its inhabitants living 'like reluctant campers, too far from home' (Pearson, 1952: 226). Nonetheless, by the 1990s, forest transitions – increases in the extent of forest cover – were beginning to occur in these places, as in many other countries (Rudel *et al.*, 2005). On Banks Peninsula, regeneration of bush has increased in coverage to between 10 and 15 per cent of the land area (Wilson, 2009; Hillary, 2011). This raises the question of whether this is an example of an emerging 'middle landscape', or a place that progresses 'both people and the land's indigenous life' (Park, 2006: 202).

To explore this, it is necessary to understand the changing dynamics of land use after the 1920s, and how this has been shaped by a liberal social order re-energized since the 1980s with the new configurations emerging from neo-liberalization (Larner *et al.*, 2007). Land use outcomes reflect changes in agricultural practices, in policies for the protection of indigenous interests and through expressions of urban amenity in a landscape adjacent to a city – Christchurch – that now has a population of 360,000 people. Until the demise of the grass seed industry in the 1930s, the work of maintaining a landscape of grass was undertaken by cocksfooters and dairy cattle. Thereafter sheep farming took over. Sheep were not compatible with cocksfoot production as their bite is sufficiently close to the ground to damage the growing part of the cocksfoot plant (Wood, 2008). But as grazing animals that do not need daily care, they are better suited to the steep slopes of the Peninsula. Most sheep are New Zealand Romneys, and well known for growing strong, white wool.

Thirty or forty years ago, Peninsula farmers received about 70 per cent of their income from wool. Today that share has fallen to nearer 15 per cent, with the emphasis now on sheep meat, some beef and supplementary sources such as tourism. This reflects a long-term decline in wool prices since the 1960s, and the removal of production and price support for agriculture with the restructuring of the state in the 1980s (Le Heron and Pawson, 1996). Farm size has increased with amalgamations of holdings. A shift towards fat lamb production has meant more intensive use of flatter land in the valley bottoms, where grass can be more readily resown and fertilized, in turn enabling lambing percentages to be raised from about 100 to 140 per cent. Better pasture supports more and faster growing lambs. And whereas formerly many farmers would cut or spray 'scrub' on the steeper slopes, today it may be left alone. Scrub is a term adopted to describe species that are invasive in pasture, both native shrubs such as *Coprosma* and kānuka, and exotics like gorse and broom.

Gorse was introduced in the 1850s as a fencing plant. The vigour with which it grows in the wetter parts of the Peninsula has long rendered farms in the southern and southeastern bays economically marginal. Increasingly, however, it is recognized both by farmers, and by town dwellers buying land, that it need not be demonized. It acts as a very effective nursery for the regeneration of native plants.

A large-scale experiment across the 1230-hectare privately owned Hinewai reserve (Figure 5.6) has demonstrated that native forest quickly returns if certain conditions, including the removal of stock, are met (Wilson, 2002). Chemical sprays halt regeneration; firing encourages the gorse seeds in the soil to grow back with renewed energy. But with pockets of bush nearby, from which birds can source and spread seed, shade-tolerant indigenous plants will grow protected by the gorse canopy. Within five to seven years, they out-compete the gorse which, requiring light, dies. This process inverts one of the shibboleths of Crosby's 'ecological imperialism', namely a greater fitness for purpose of introduced northern hemisphere species.

This new perspective on gorse, alongside the changes in agricultural production, goes some way to explaining the resurgence of the indigenous on the Peninsula. In turn, this is shaping, unevenly, but clearly, the emergence of different values in landscape, in which what was once seen as 'unimproved' is no longer demeaned as 'waste'. Evidence for this is the extent to which landholders have adopted covenanting schemes that provide a legal mechanism to secure in perpetuity features of their land that they wish to see protected and retained beyond their tenure. This represents a new form of enclosure, or more accurately 'exclosure', since stock-proof fencing is required and public access is unusual. The Queen Elizabeth II National Trust has been active on the Peninsula since the early 1980s, and has about 70 registered covenants. The Banks Peninsula Conservation Trust, established in 2001 by a coalition of landholders, has over 50 (Figure 5.6). Its covenants are mainly on farms and protect areas of high biodiversity and open spaces such as headlands. It was set up to counter attempts by local government in the late 1990s to override private property rights by regulating for protection (Hillary, 2011).

In contrast, less than one-third of the Queen Elizabeth II Trust's covenanters are farmers. The rest own residential lifestyle blocks, holiday homes, or small conservation blocks without homes, pointing to the impact of urban amenity in the landscape. In all, the extent of covenanted land is small, but it sits alongside publicly owned categories of conservation land, which are vested in and managed by the Christchurch City Council or the Department of Conservation (Figure 5.6). The Scenery Preservation Act of 1903 was the first dedicated state measure for the reservation of native nature in New Zealand; its provisions were early used to set up a number of scenic reserves on the Port Hills, the Peninsula's northern flank, and southward along the ridgeline towards Akaroa. The Act was the result of the drive of a Member of the House of Representatives for Christchurch, Harry Ell, who was motivated by a desire to encourage access to the countryside for urban people. The publicly owned reserves (and Hinewai) allow for this. The majority of the Peninsula, however, is open only to gaze upon as scenery, otherwise locked away as an enclosed landscape lacking even the public rights of way that remain unextinguished in much of Europe.

The historical ecologist Geoff Park characterized New Zealand as 'a land of two kinds of country: one in which the urge was to advance human activity, and

FIGURE 5.6
Map of
protected areas
on Banks
Peninsula in
2013, including
public reserves
vested in or
managed by the
Christchurch
City Council
and the
Department of
Conservation,
covenants on
private land,
and the
Hinewai private
reserve.

Source: Based on
data supplied by
the Christchurch
City Council and
the Department
of Conservation,
and the map in
Hillary (2011:
59).

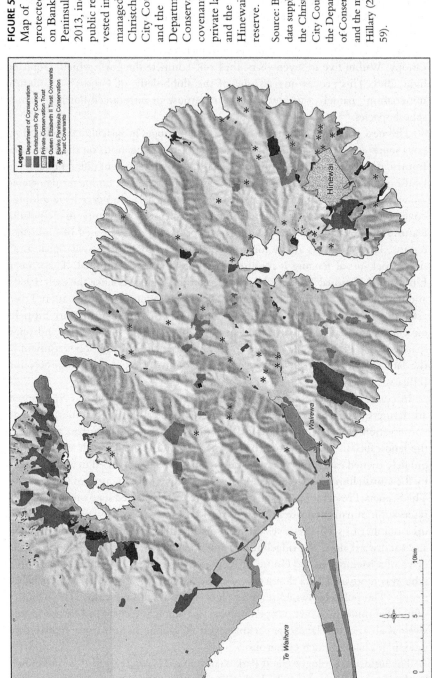

another in which the urge was to *exclude* it' (Park, 2006: 202, emphasis in original). The 30 per cent of the national land area that is occupied by the conservation estate, being national parks and lesser categories of conservation land, comprises the second category. It is there that the most extensive tracts of indigenous vegetation survive. A middle landscape is one that has some attributes of both kinds of country. On Banks Peninsula, the emergent mosaic of native bush patches and corridors across farmland encourages the return of birdlife, with growing populations of kereru (native pigeons) and bellbirds. Tui have been successfully reintroduced by the Banks Peninsula Conservation Trust (Hillary, 2011). Officially protected areas may only provide spot coverage in the landscape, but between them is a growing skein of bush life that constitutes a public good for all who look and listen. In this sense, the once-tidy clarity of the Peninsula's landscape of grass is being modified in ways that begin to blur the boundaries of private and public.

If space is made for displaced native nature, then a middle landscape also creates space for native people. The Ngāi Tahu Claims Settlement Act 1998 made economic redress to Ngāi Tahu for property rights lost through dispossession. Some of this has been used to underwrite its 18 regional *papatipu rūnanga* (local councils), four of which have *marae*, or meeting houses, on the Peninsula. The Act also returned ownership of the bed of Te Waihora/Lake Ellesmere to Ngāi Tahu, given its importance as a food basket noted for eels, flounder and waterfowl (Figure 5.6). A co-governance agreement has since been developed by Te Waihora Management Board between Ngāi Tahu and the Canterbury Regional Council, along with a joint management plan with the Department of Conservation for the lake bed and adjacent conservation lands (Stevens, 2013). Nonetheless the lake is one of the most polluted in the country, being a shallow sump into which drains nitrate-enriched run-off from surrounding farms. The extent of change can therefore be exaggerated, but Park's claim that there is 'No framework for living *with* the indigenous, in any other form than visiting and admiring it' (Park, 2006: 202, emphasis in original) is less the case than even a few years ago.

Conclusion

The question that this chapter posed at the outset was how have we exercised our 'gargantuan agency' and how we might come to terms with 'an almost unbearable level of responsibility' in the Anthropocene (Gibson-Graham and Roelvink, 2009: 321)? The narratives discussed here have provided some insight into these two issues. Banks Peninsula typifies one of the great landscape changes of the Anthropocene that occurred with the incorporation of remote places within circuits of imperial expansion. This was the intentional displacement of indigenous vegetation, in this case 'the bush', by introduced 'English grasses' and their fellow travellers such as gorse. Territory was appropriated cheaply as 'wastelands' from its native occupants. Monetary value was created through enclosure, enabling the benefits of 'improvement' to be captured by new owners. An essential part of improvement was the use of imported biota to seed and contain new fields of production. In turn the Peninsula became an important source of seed for the repair and extension of pastures elsewhere, showing how in the Anthropocene flows of biota have criss-crossed in tangled webs.

Landscaping has been the product of, and shaped the labour of, both humans and non-humans: sawyers, farmers, cocksfooters, townsfolk, cattle and sheep. The agency of plants, birds and trees has become increasingly significant as a landscape of grass has given way to a more complex assemblage of pasture, resurgent bush and settlement. The picture that Charles Elton drew half a century ago of New Zealand exposed to 'a steady stream of aggressive invaders' (Elton, 1958: 59) needs to be modified. People are now working with the attributes of invaders like gorse to enable native and introduced birds to re-establish both native flora and fauna. At the same time vigorous programmes are being pursued to kill other invaders that undermine this objective, animals such as possums, wild goats, mustelids and rats, and plants like old man's beard, clematis and sycamore (Schmechel, 2009). This new accommodation is an example of what Elton himself described as 'looking for some wise principle of co-existence between man and nature, even if it has to be a modified kind of man and a modified kind of nature' (Elton, 1958: 145). The ways in which spaces are being created for living with the indigenous represent a conscious re-alignment of human interests away from the prioritization of environmental dominion and towards recognition of our connections with landscape and history, as well as our dependence on nature.

Recently invasion biologists have debated vigorously about whether or not we should learn to live with 'novel ecologies' (Davis *et al.*, 2011; Simberloff, 2011). It has been suggested that an embrace of novelty might enable us to put edenic visions aside and 'dispense with imaginary places to which there is no hope of return' (Robbins and Moore, 2012: 16). Others have begun to argue for programmes of 'rewilding', in which the interests of the introduced *and* the indigenous, of farming *and* of native ecosystems are recognized (Monbiot, 2013). At least part of an answer as to how we might assume greater responsibility for our actions in nature is through a growing acceptance of messy, emergent 'middle landscapes' in which there is clearer agreement of how to intervene and to what ends. The argument of this chapter has been that these middle landscapes represent one response to the fretfulness of the Anthropocene, and that New Zealand's Banks Peninsula is one example of what a middle landscape might look like.

Acknowledgements

We are grateful to Edward Aitken and Paul de Latour for talking about Banks Peninsula farming, and to Alice Shanks, the Queen Elizabeth II National Trust representative for the Peninsula. Garth Cant, Ian Hodge, Harvey Perkins, Stig Roar Svenningsen and Vaughan Wood kindly discussed aspects of the draft with us.

References

Ballantyne, T. 2012. *Webs of Empire: Locating New Zealand's Colonial Past*, Wellington, Bridget Williams Books.

Barry, J. 1999. *Environment and Social Theory*, London, Routledge.

Belich, J. 1996. *Making Peoples: A History of the New Zealanders: From Polynesian Settlement to the End of the Nineteenth Century*, Auckland, Allen Lane.

Brooking, T. and Pawson, E. 2011. *Seeds of Empire: The Environmental Transformation of New Zealand*, London, I.B. Tauris.

Christensen, A. A. 2013. Mastering the Land: Mapping and Metrologies in Aotearoa New Zealand. In: Pawson, E. and Brooking, T. (eds) *Making the Land: Environmental Histories of New Zealand*, new edn, Dunedin, Otago University Press, pp. 310–327.

Clark, A. H. 1949. *The Invasion of New Zealand by People, Plants and Animals: The South Island*, New Brunswick, NJ, Rutgers University Press.

Clout, H. 1998. Rural Europe Since 1500: Areas of Innovation and Change. In: Butlin, R. A. and Dodgshon, R. A. (eds) *An Historical Geography of Europe*, Oxford, Clarendon Press, pp. 225–242.

Crosby, A. W. 1986. *Ecological Imperialism: The Biological Expansion of Europe, 900–1900*, Cambridge, Cambridge University Press.

Cumberland, K. B. 1941. A Century's Change: Natural to Cultural Vegetation in New Zealand, *Geographical Review*, 31(4), 529–554.

Davis, M. A., Carroll, S. P., Thompson, K., Pickett, S. T. A., Stromberg, J. C., Del Tredici, P., *et al.* 2011. Don't Judge Species on Their Origins. *Nature*, 474, 153–154.

Elton, C. S. 1958. *The Ecology of Invasions by Animals and Plants*, London, Chapman and Hall.

Fairburn, M. 2012. Edward Gibbon Wakefield. In: *Te Ara: The Encyclopedia of New Zealand*. Available at: www.TeAra.govt.nz/en/biographies/1w4/wakefield-edward-gibbon.

Gibson-Graham, J. K. and Roelvink, G. 2009. An Economic Ethics for the Anthropocene. *Antipode*, 41, 320–346.

Goldstein, J. 2013. *Terra Economica*: Waste and the Production of Enclosed Nature. *Antipode*, 45, 357–375.

Harvey, D. 2003. *The New Imperialism*, Oxford, Oxford University Press.,

Hillary, N. (ed.) 2011. *Banks Peninsula Conservation Trust: Ten Years of Protecting Biodiversity*, Akaroa, Banks Peninsula Conservation Trust.

Ingold, T. 2000. *The Perception of the Environment: Essays on Livelihood, Dwelling and Skill*, London: Routledge.

Jacobson, H. C. 1893. *Tales of Banks Peninsula*, 2nd edn, Akaroa, *Akaroa Mail*.

Johnston, J. A. 1981. The New Zealand Bush: Early Assessments of Vegetation. *New Zealand Geographer*, 37, 19–24.

Larner, W., Le Heron, R. and Lewis, N. 2007. Co-constituting 'after Neoliberalism': Political Projects and Globalizing Governmentalities in Aotearoa/New Zealand. In: England, K. and Ward, K. (eds) *Neoliberalization: States, Networks, Peoples*, Malden, MA, Blackwell Publishing, pp. 223–247.

Le Heron, R. and Pawson, E. (eds) 1996. *Changing Places: New Zealand in the Nineties*, Auckland, Longman Paul.

Lester, A. 2006. Imperial Circuits and Networks: Geographies of the British Empire. *History Compass*, 4, 124–141.

Livingstone, D. N. 2005. Science, Text and Space: Thoughts on the Geography of Reading. *Transactions of the Institute of British Geographers*, New Series, 30, 391–401.

Marx, K. 1976. *Capital: A Critique of Political Economy*, vol. I, Harmondsworth, Penguin Books.

McCarthy, J. and Prudham, S. 2004. Neoliberal Nature and the Nature of Neoliberalism. *Geoforum*, 35, 275–283.

Mitchell, W. J. T. 1994. Imperial Landscape. In: Mitchell, W. J. T. (ed.) *Landscape and Power*, Chicago, University of Chicago Press, pp. 5–34.

Monbiot, G. 2013. *Feral: Searching for Enchantment on the Frontiers of Rewilding*, London, Allen Lane.

Olwig, K. R. 2008. Performing on the Landscape Versus Doing Landscape: Perambulatory Practice, Sight and Sense of Belonging. In: Ingold, T. and Vergunst, J. L. (eds) *Ways of Walking: Ethnography and Practice on Foot*, Farnham, Ashgate, pp. 81–91.

Park, G. 2006. *Theatre Country: Essays on Landscape and Whenua*, Wellington, Victoria University Press.

Pawson, E. 2008. Plants, Mobilities, Landscapes: Environmental Histories of Botanical Exchange. *Geography Compass*, 2, 1464–1477.

Pearson, B. 1952. Fretful sleepers. *Landfall*, 6, 201–230.

Petrie, L. M. 1963. From Bush to Cocksfoot: An Essay on the Destruction of the Banks Peninsula Forests, MSc thesis, University of Canterbury, NZ.

Pickles, J. 2004. *A History of Spaces: Cartographic Reason, Mapping and the Geo-coded World*, London, Routledge.

Potts, T. H. 1868. The Forests of the Colony. *New Zealand Parliamentary Debates*, vol. 4, 25 September to 20 October, pp. 188–189.

Robbins, P. and Moore, S. A. 2012. Ecological Anxiety Disorder: Diagnosing the Politics of the Anthropocene. *Cultural Geographies*, 12, 3–19.

Rudel, T. K., Coomes, O. T., Moran, E., Achard, F., Angelsen, A., Xu, J. and Lambvin, E. 2005. Forest Transitions: Towards a Global Understanding of Land Use Change. *Global Environmental Change*, 15, 23–31.

Sauer, C. O. 1963. The Morphology of Landscape. In: Leighly, J. (ed.) *Land and Life: A Selection from the Writings of Carl Ortwin Sauer*, Berkeley, CA, University of California Press, pp. 315–350.

Schmechel, F. 2009. *The War on Pests: Dealing with Key Pest Plants and Animals that Threaten Native Species: A Landowner's Guide for Banks Peninsula and Kaitorete Spit*, Christchurch, Banks Peninsula Conservation Trust and Environment, Canterbury, NZ.

Simberloff, D. 2011. Non-Natives: 141 Scientists Object. *Nature*, 475, 36.

Stevens, M. J. 2013. Ngāi Tahu and the 'Nature' of Māori Modernity. In: Pawson, E. and Brooking, T. (eds) *Making the Land: Environmental Histories of New Zealand*, new edn, Dunedin, Otago University Press, pp. 293–309.

Waitangi Tribunal. 1991. *The Ngāi Tahu Report 1991 (Wai 27)*, vol. 2, Wellington, Brooker and Friend.

Wilson, H. 2002. *Hinewai: The Journal of a New Zealand Naturalist*, Christchurch, Shoal Bay Press.

Wilson, H. 2008. Vegetation of Banks Peninsula. In: Winterbourn, M., Knox, G., Burrows, C. and Marsden, I. (eds) *The Natural History of Canterbury*, 3rd edn, Christchurch, Canterbury University Press, pp. 251–278.

Wilson, H. 2009. *Natural History of Banks Peninsula*, Christchurch: Canterbury University Press.

Wood, V. 2008. Akaroa Cocksfoot: Examining the Supply Chain of a Defunct New Zealand Agricultural Export. In: Stringer. C. and Le Heron, R. (eds) *Agri-food Commodity Chains and Globalising Networks*, Aldershot: Ashgate Publishing, pp. 216–227.

Wood, V. and Pawson, E. 2008. The Banks Peninsula Forests and Akaroa Cocksfooting: Explaining a New Zealand Forest Transition. *Environment and History*, 14, 449–468.

Wylie, J. 2011. Landscape. In: Agnew, J. A. and Livingstone, D. N. (eds) *The SAGE Handbook of Geographical Knowledge*, London, Sage Publications, pp. 300–315.

Wynn, G. 1979. Pioneers, Politicians and the Conservation of Forests in Early New Zealand. *Journal of Historical Geography*, 5, 171–188.

Wynn, G. 2013. Destruction under the Guise of Improvement? The Forest, 1840–1920. In: Pawson, E. and Brooking, T. (eds) *Making the Land: Environmental Histories of New Zealand*, new edn, Dunedin, Otago University Press, pp. 122–138.

Zalasiewicz, J., Williams, M., Haywood, A. and Ellis, M. 2008. Are We Now Living in the Anthropocene? *GSA Today*, 18, 4–8.

PART III

Everyday life in invasion ecologies

6

LIVING IN A WEEDY FUTURE

Insights from the garden

Lesley Head

The garden as study site – and what it says about nature

On a hill close to the centre of Wollongong, New South Wales, remnant stands of spotted gum (*Eucalyptus maculata*) are protected in a small formal reserve and in some private gardens. Three different householders living on different sides of that hill represent some of the variability in how Australians understand and experience gardens, nature and weeds. Lennie and Connie's backyard was dominated by an extensive vegetable garden and chook shed maintaining traditions they brought from Italy more than 40 years previously. The productive vegetable garden was meticulously weeded. Lennie had established some small vegetable beds on the adjacent reserve, where he also grazed his rabbits in their mobile hutch. He was very careful to protect spotted gum seedlings, which he marked with stakes and tape, and was in active discussions with the local council officers about these activities. In talking about his garden, Lennie did not talk about endangered species but rather about productivity and his family and being involved with the soil. Nevertheless the outcome was ongoing stewardship of a locally endangered species.

On the other side of the hill lived Kris, an environmental scientist. The remnant stand of *E. maculata* and other eucalypts was the reason she bought her block, which contained a number of very large spotted gums. She had been actively trying to restore the native vegetation since moving in, by removing 'lawn, trees, azaleas, geraniums . . . and a whole lot of other pests'.

Further down the hill, in Mira's backyard, the strongest impression for the visitor was of order and tidiness. Mira described this area as being like a 'small house', which it was necessary to look after, clean and decorate. When she mowed her lawn or fed her roses, she was loving and nurturing a garden which was 'everything in my heart'. Despite, or perhaps because of, her demanding full-time job, her morning routine began with half an hour in the garden. She described this as a time that 'makes me relaxed', when she noted the cycles of plants and their flowering, pulled weeds and planned what needed further tidying.

According to many theories about settler Australian environmental relations, Mira was alienated from nature through taming and domesticating it, and Lennie projected a European ethic onto it, rather than coming to terms with the essence of Australian nature. Kris's backyard work would be seen as representing the appropriate conservationist response, but because it was done in an industrial city, it would be deemed far less important than her professional work in nature protection outside the city. All three backyards would be deemed peripheral to the urgent work of protecting the 'real' nature in remote areas. I have argued previously that such perspectives contain two central rifts; between an immigrant Australian nation and its environment, and between nature and the city (Head and Muir, 2007). For more than two hundred years in Australia, the distinction between the city and the bush has roughly paralleled that between culture and nature. When we preserve nature, we preserve it 'out there'.

In this chapter I revisit the backyard garden study to see what it can add to current debates about invasive plants and weeds. The connection between gardens, agriculture and weeds throughout human history is well known (Harlan and De Vet, 1975). In the Australian context we know that many of our most problematic weeds have 'jumped the garden fence' (Groves et al., 2005). They were brought into the country for their beauty, and to remind migrants of home, but have behaved differently in Australian conditions, often becoming invasive.

The garden focus of this chapter is not because that will teach us how to deal with weeds in 'pure' nature somewhere else. It is not to help us get things back inside the garden fence. Rather, the premise of the chapter is that if we cannot live without weeds, the garden is (just) one of the spaces where we can understand the complexity and diversity of what it means to live with them. Whatever our preferences, all the evidence suggests that we are likely to live with more weeds in future than less. Weeds will become more numerous, abundant and widespread. Their influence will be geographically variable and they will interact with humans and other organisms in unpredictable ways. It is increasingly acknowledged that Invasive Plant Management (IPM), though a significant global issue, is a matter of coexistence rather than control. Despite these contingencies, management and legislative rhetoric around weeds are often framed in less flexible terms, using themes such as the 'war on weeds', with metaphors of invasion, competition and war being pervasive (Larson, 2008; Downey, 2011). Such framings 'naturalise antagonistic ways of relating to the natural world' (Larson, 2008: 169). Thus science-based policy can tend to see its work as educating the community in a linear direction about appropriate management of invasive plants. And whole industries are built around killing weeds, in contexts from the garden to the farm to the national park.

A further contextual statement is that mainstream Australian perspectives on nature focus on the past rather than the present or future. Yet all the scientific information we have is that profound thresholds have been crossed, and restoration of past environmental conditions is not possible. For a combination of reasons – encompassing land use change, extinctions, climate change, and social changes – our future will look very different. Weeds are now integrated into contemporary ecosystems, becoming part of what Richard Hobbs and colleagues (2006) have called 'novel ecosystems' or what Tim Low (2002) calls 'the new nature'. Simply removing them will not restore a previous set of conditions; there is no going back. This does not mean 'giving up' on weeds, but rather thinking more carefully about when and where humans can and should intervene (Larson, 2008).

Nor do I want to suggest that the past was ever somehow 'pure' nature. Weeds have been our companions throughout human history and always will be. The broader issue is not just how to live in a weedy present and future, but how to approach a future that is open, inherently uncertain and subject to processes only partly under human control.

I want to suggest that we use weeds to help us understand the contradictions inherent in being humans – in and of nature. Weeds, like we humans, occupy ambiguous space in thinking about nature. We can use this ambiguity to think with. Different patterns of human ecological practice – growing crops, making roads, leaving city blocks derelict – encourage different combinations of weeds.

This chapter approaches the issue from the opposite direction to the management directive to get rid of weeds, drawing on vernacular experience of and engagements with weeds and invasive plants. It asks, what can we learn from people who live with weeds in different everyday contexts? The examples presented here are drawn from a project on suburban gardens; in other work we are also examining the everyday experience of professional weed managers and policy-makers (Atchison and Head, 2013). I do not suggest that learning to live with weeds is a comfortable process, as the discussion of agency demonstrates below, nor one without hard decisions between different kinds of environmental damage. Living with weeds often necessitates killing them.

The agency of plants

There has been considerable recent discussion in cultural geography and associated disciplines about the agency of nonhumans, or more specifically the complicated combinations of agency that produce the world (Haraway, 2008). Attention to animal agency has recently been extended to plants, in contexts that include trees (Jones and Cloke, 2002), gardens (Hitchings, 2003; Power, 2005), invasion (Barker, 2008; Ginn, 2008), crops (Head et al., 2012) and seeds (Phillips, 2013). A number of studies have analysed vernacular human experiences of the agency of plants (Cloke and Jones, 2001; Hitchings, 2003; Hitchings and Jones, 2004; Power, 2005). Such experiences are characterised by people in both positive and negative terms. Plants display individual liveliness and beauty (Hitchings, 2003), exert calming influences (Hitchings, 2006), and plants and people draw one another into patterns of care (Power, 2005). On the other hand, plants have lives of their own beyond human control, which can lead to uneasiness and 'awkward encounters' when 'plants are perhaps no longer often thought about in terms of their capacities and behaviours' (Hitchings, 2007: 372). For Ginn (2008), animals and plants are both active participants and subversive agents in the colonial landscape of Aotearoa New Zealand, the context in which Barker (2008) discusses contemporary agency and changeability of gorse and its management.

The relevance of considering plant agency in more detail in this chapter is that if we want to understand weedy landscapes as complex co-productions between human and other causes, it is important to consider how people attribute agency in their lives with weeds. It is my intention here to illustrate how and in what circumstances gardeners attribute agency to plants.

The backyard garden study

The original project was a large study of 265 backyard gardens and their 330 owners (some couples and family groups participated). Most of the fieldwork was carried out between 2002 and 2005, a period of intensifying drought in southeastern Australia. The sample population comprised 122 backyards in Sydney, 119 in Wollongong, and 24 in Alice Springs, in areas that spanned the socioeconomic, demographic and ecological variability of each city area. About 90 per cent of the sample were households with detached house and garden, the remaining 10 per cent being apartment or unit dwellers with small courtyards or balconies. For further details of the study and its methods, see Head and Muir (2004), Head and Muir (2006), and Head and Muir (2007). Diverse types of engagement with nature and the garden were sought in the study; some backyarders identified as passionate gardeners, while others strongly identified as non-gardeners. As background it is relevant that the study population bears out other research showing that the most popular types of garden in Australia include exotic plant species, alone or in combination with natives (NPWS, 2002; Zagorski et al., 2004; Trigger and Mulcock, 2005). Committed native gardeners – the ones we called the 'purists' – are a demonstrably minority group. Most participants were comfortable with the hybrid reality of combinations of native and exotic plantings, with affection for their different attributes.

The agency of weeds and invasives

Weeds are commonly defined as plants that are in the wrong place, and that do not belong because they are not native. A large body of scholarship now shows that concepts such as 'nativeness' tell us more about human categories than anything inherent in the plants or their evolutionary processes (Gould, 1997; Bean, 2007; Davis et al., 2011; Head, 2012). So the point of my argument is not to focus on different definitions of what belongs or does not, but rather to examine the range of practices towards weeds. Backyard gardeners like Lennie, Kris and Mira each define weeds differently, but their practices have certain things in common. More broadly, I will show that some practices towards weeds match up with definitional attitudes towards native or non-native plants; others cross-cut them. In practice, weed managers of all sorts separate ideas of belonging (what plants are) from behaviour (what plants do). We attend to invasive plants – in our gardens, in our crops – and leave the ones that behave themselves alone.

This chapter uses the terminology of *weed* rather than *invasive*, as it resonated more strongly with the study participants, and emerged during our interview conversations. Most participants were much more likely to talk about weeds in terms of their invasive qualities than their nativeness *per se*. Indeed, for them, a weed is in practice a plant that invades. This is a common-sense understanding borne out of the labour of maintaining a garden, or frustration at not having the time or inclination to do so. Thus Angus talked of weeds as plants that 'run over everything else'. For Jo, an active participant in bush regeneration projects, there was a strong distinction between good and bad exotic plants, separate to their nativeness. The bad ones, including Madeira vine (*Anredera cordifolia*) and lantana (*Lantana camara*), are invasive in the bush, while those that sat quietly in the domestic space of her garden (specimen conifers, port wine magnolia, daffodils) are very welcome despite being exotic. Invasive plants were thought of as exerting agency much more commonly than other plants; the agency of weeds is clear, but unwelcome.

Further support for this vernacular view of agency was provided by an in-depth linguistic analysis on a subset of three interviews from different gardener types. The findings show the conditions under which the agency of animals and plants was understood by our human participants (Thomson *et al.*, 2006). Animals and plants were construed as Agents when perceived as the cause of an unwelcome process or else undergoing something 'natural'. For example, the quotes, 'They [the magpies] (Agent) are going to go for us' and 'They [weeds] (Agent) make it look more mottled . . .' indicate unwelcome, agentive processes by magpies and weeds. The quote, 'They (Agent) bear fruit' is an example of a 'natural' agentive process by a tree. In contrast, non-weedy plants were expressed as places or locations where things happen, as in 'Actually people drive on the front lawn' and '. . . and all the undergrowth around them [roses and lilli-pillis]'. In general in these interviews, humans are the agents of what goes on. Animals and plants undergo processes or else just 'are' or 'happen'. However, if plants, animals and the physical elements are involved in agency, then typically the impact of their agency is construed as negative. In this way they are held responsible for their action.

The fact that the inanimate world is construed as positive only when it 'behaves' itself is consistent with the frequent concern in the broader sample over matter (e.g. trees, weeds, mess) being 'out of place'. It is tied up with the gardeners' reasons or motivations behind their gardening practices, and their understanding of what belongs in different places.

Practices with weeds

In this section I outline the main practices of weeding – albeit these are overlapping categories – that emerged in interviews, observations and garden walks.

Weeding as tidying

For perhaps the majority of people, but for different reasons, neatness and tidiness are associated with order, beauty and happiness. There are various ways by which tidiness can be achieved, and indeed a variety of perceptions about what constitutes tidiness. For example, someone who loves mowing lawns, and a passionate native gardener who removes exotics, are both 'tidying up' in the sense of creating order in their environment. This engagement can be summarised in the words of the participant who said, 'As long as it's tidy, I'm happy', and another who said, 'When it's tidy, you feel better.'

Conversely, lack of tidiness can induce shame; 'I'm ashamed of it now because it's a bit of a mess.' The garden cannot perform its role as a place of peace or leisure if it is a mess, and will become rather a place of stress (another set of demands in already busy lives) and work (or guilt, if the work is not done). These emotions were particularly experienced by women, such as the one who said, 'It irritates me if I see it looking too scruffy.' Another, a working mother, said wearily of her garden, as if of another child, 'I resented the mess and the constant need.'

With a few exceptions, we interpreted tidying practices not as the 'domination of nature' often critiqued within the Western psyche, but rather as an expression of a need for order within busy lives and dynamic domestic spaces. Most people wanted a living environment that was clean, safe and tidy, and the reason nature and weeds should stay 'out there' is because of their challenge to those attributes, usually expressed in terms of dirt, danger and/or mess. Tidiness was valued for a complex set of reasons that included social respectability, a certain moral quality, and the stress occasioned by mess. Weeding as tidying recognised the importance of nature being able to 'do its own thing', indeed, it is often able to celebrate that, provided it is not right on the doorstep. It is also a spatial practice; most people wanted to put some distance between themselves and the perceived messiness of nature.

The labour of weeding

Thus tidying practices show both negative and positive dimensions, such as those associated with work and guilt. Like housework, there is always weeding to do, as Quentin recounted:

> Mainly at the moment I would get out there and actually weed it and tidy it up again. It actually looked good at one time but I've just left it because I've been on shift work and just so busy lately that I haven't had time to actually do anything, so it's just gone on to how it has at the moment.
>
> *(Quentin, Albion Park)*

The labour of initial weed removal is different from maintenance. Various study participants described the huge task to remove lantana – including with goats and bulldozers – when they moved into their house and garden. The maintenance task is much more painstaking and requires ongoing, habitual commitment, as described by Jo, the bush regenerator:

> Now where that fern is, all that area was all covered in madeira vine and lantana, and so I've just gradually got rid of it all and I have to keep vigilant. The madeira vine still comes back but we've just worked on, I just work at it, you know, it's painstaking work but I suppose I seem to have the right kind of mentality . . . when something gets away, you just go back into one area and clean that completely . . .
>
> *(Jo, northern Wollongong)*

The issues that Jo was dealing with on an ongoing basis were very similar to those encountered by professional weed managers in more remote areas, such as those we have worked with in the Northern Territory (Atchison and Head, 2013). For example, Dennis's experience, if somewhat more aggressive than Jo's, was very similar:

> I'm aware that I'm bashing my head against a brick wall. I killed millions yesterday. Millions more will grow in their place, but I can't wait to get back there and kill all them ones as well.
>
> *(Dennis, weed manager WA, quoted in Atchison and Head, 2013)*

It was quite rare for people to say they enjoyed the work of weeding in the garden. Margaret of Campbelltown provided one example: 'I sometimes think weeding and tidying up is work. But once I get into it I really like it. I like puddling around in the dirt.'

Weeding for vegetable production

Much of what we know about the history of weeds is intertwined with the history of agriculture, which has involved concentrating on some highly productive food plants at the expense of others. Like broad-scale agriculture, producing vegetables in backyard gardens usually entails a commitment to removing weeds. (The exception to this is some permaculture gardeners who are happy to let everything go.)

> I spend time out here growing things, coming out here every day and seeing where the weeds are, checking how the vegetables are. I don't think you can grow vegetables if you don't spend a lot of time checking how they are.
>
> *(Karen, Wollongong)*

> I always try to keep the soil clean around the plant because if one [weed] is allowed to start growing between the plants, it takes all the goodness from the soil and from the plant.
>
> *(Blaga, Port Kembla)*

In contrast to the general labour of removing weeds in the previous section, the labour of removing weeds so that vegetables can grow did produce satisfaction and pleasure. But that pleasure was usually expressed in relation to the tactility of the soil, and the growth of the favoured plants, rather than with reference to the weeds themselves.

Weeding as purification

Kris, whom we met in the introduction, was an example of a group we called committed native gardeners or purists. This group tended to express disparaging attitudes towards 'exotic' or 'foreign' plants, and the neighbours who enjoyed them. For them, the importance of natives as 'belonging' was paramount. For some in this group the notion of gardening, with its connotations of planting and humanly assisted productivity, was inaccurate. Rather they saw themselves as restorers of native bushland, or eradicators of weeds that prevent native bushland restoring itself.

Claire in Alice Springs had been on a mission for many years to remove the introduced buffel grass (*Cenchrus ciliaris*) from her family's block:

> My object is to eradicate all the buffel and noxious grasses out and let the natural regrowth occur and it's just, it's my art; we have a sense that if I can achieve that to a high degree, that, that's my art within the community.

Soon after rain is an important time for the removal of buffel, when it rapidly responds by creating a green sward across the landscape. Claire described the practice of her art in very embodied terms. She told of regularly starting at 4.30 or 5.00 on summer mornings and removing buffel for three hours before the business of the day began, coming inside drenched in sweat and satisfied with her exertions. This physical, almost sensual, engagement and investment of labour were similar to that described by passionate gardeners who were planting things, and lawn mowers who were removing things.

Weeding the bush: the practice of spatial purification

Many gardeners, and not just purists, were active in weed removal beyond their fences, seeing different types of nature as belonging in different places. This is a further example of weeding as a *spatial* purification. Claire was not the only Alice Springs resident who removed buffel. Chris, who had an extremely manicured backyard including lawn and rose beds, was active in weed removal beyond her open mesh fence, seeing different types of nature as belonging in different places.

In Wollongong, Janine's backyard garden was conceptually similar to Chris's, albeit with different species. A fence separated the more domestic part of her backyard from the adjacent bush. On the inside were grass, vegetables, garden beds and homes for her extensive menagerie of pets. On the 'outside' of the fence, but still on Janine's land, was an area that she was regenerating, extending down to the creek.

> The inside of the backyard I've got a mixture . . . but on the other side of the fence everything that I plant out there is, like, local and what belongs there. There's actually three old camellia trees out there that are quite big which I'm going to have to cut down because they just don't belong there.

Janine would have no qualms about leaving the exotic camellias if they were a few metres away, inside the fence.

In a bushland suburb on Sydney's leafy north shore, Coralie walked often in the National Park at the end of her street. She actively pulled out weeds there with her friends and she lobbied the council about adverse growth on disturbed ground in the park. In a nearby suburb, Don described his activities in the surrounding bush:

> I sort of look after the end of my street which is part of the Thornleigh bush park area. It covers a number of streets including Pennant Hills and I do that once a month for about three hours. That involves weeding and identifying the plants and getting a lot of things done; trying to get the natives back to what it was before.

Weeding the bush can work in two ways. Coralie and Dan exemplify the practice of removing non-native weeds from bushland. Conversely, Lennie in our Introduction, and another couple who have an open boundary with a national park exemplify a kind of cleaning and tidying practice out onto adjacent bushland. As the husband said, pointing to a mown area outside his fence, 'That's part of our garden.' And his wife continued, 'Yes, he's been pruning the natives out there and they do come good. They've got the straggly.'

Conflict over weeds: lawn as example

Nothing better exemplifies the conflicting attitudes over weeds than lawn. There was a strong polarity between those who loved lawn and those who hated it. Lawn lovers focused on the sensory pleasures of grass, the pleasure of the labour of mowing and lawn's importance as a play area for children. The desire for lawn was often related to a desire for control and order. Lawn areas were kept pristine and weed free, with some participants talking about crawling around on their hands and knees removing broad-leaved plants that marred their green oasis.

In contrast, for lawn haters, the lawn itself is the weed. They voiced their sentiments in terms of excessive water used, condemning lawns as environmentally unsustainable in the Australian environment. Kris described herself as being on a 'mission' against lawn. The exception was the commonly expressed feeling that kids need lawn. So lawn haters sometimes qualified their antipathy with the statement that they did not have young children at the moment.

Trees as weeds

There was some sense that well-established trees cannot be weeds. For example the things that make camphor laurel (*Cinnamomum camphora*) a successful weed also create a spectacular tree. In one household, a whole entertaining deck had been built around a well-loved camphor laurel, its use changing with the growth of the family. The social life of the tree overrode its weed status in terms of belonging.

There seems to be something about the size, life form and longevity of trees that leads people to ascribe more rights to them than other forms of life. Even when people saw trees as messy and potentially dangerous, many did not like cutting them down. Barbara described the clearing they had to do when she and her husband came to their house as 'quite heart-breaking, especially for Brian, cutting down trees [camphor laurels] that were weeds'.

Weeding as never finished

There was widespread recognition among study participants of the dynamism and constant change in the nonhuman world. Maggie, from inner Sydney, expressed this view that nature has a life of its own beyond human control:

> A garden has a sense of discovery and it has a sense of relationship, that you are having a relationship with nature; in a sense it's not yours, you are the keeper of it in a sense and things are revealed to you, nice surprises and nasty surprises like something is eating my lovely orchids.

For gardeners, the non-gardener notion that the garden could ever be 'finished' would be incomprehensible. As one said, 'Well, I think the thing about a garden is that it should change.'

> Gardens are never finished, I don't think. They are kind of like art, like a lot of people describe paintings as they are never finished . . . a garden is a constant work in progress.
>
> *(Amber, inner Sydney)*

> Well, for me, it's kind of a work in progress and things will gradually change.
> *(Shooshi, Wollongong)*

Conclusion: towards a more open future

In this chapter I have sought to examine the weeding practices of backyard gardeners in order to help us think more carefully about what it takes to live with weeds. These gardeners have different definitions of weeds, and a range of attitudes to specific plants and where they belong. This leads to diverse, contradictory and sometimes conflictual practices. To conclude, I focus here on the similarities across the sample that might inform wider debates about invasive plants.

First, in this study, plants were most commonly construed as agentive when doing something unwanted. As humans, we seem to have least trouble acknowledging plant agency when it is negative. Second, it follows that unwanted plant agency invokes many different types of human response, summarised here as a range of practices including tidying, production, and purification. These weedy practices provide a basis for comparison with other studies in different natural resource management contexts. They are also spatially uneven, depending on where people understand certain types of nature to belong (cf. Gill *et al.*, 2010).

Third, weed management takes a lot of effort (whether pleasurable or not), invested over the long term. Initial investment and maintenance require different kinds of labour and vigilance. It is a job that will never be finished, but carefully targeted, long-term labour can make a difference. The most successful and contented garden weed managers are comfortable with the fact that plants have a life of their own and do their own thing.

Finally, I argue that we need a more open acknowledgement of the contradictions, edginess and difficult choices that attend contemporary Australia's engagements with nature. The times require us to go beyond the ideal of a pristine past and more honestly face a fraught, unpredictable and surprising future. Resilient, opportunistic, larrikin weeds may be more useful companions on that journey than we can yet imagine – and gardeners who live with them one of our instructors.

Acknowledgements

Thanks to the Australian Research Council for funding (DP0211327 and FL0992397).

References

Atchison, J. and Head, L. 2013. Eradicating bodies in invasive plant management. *Environment and Planning D. Society and Space*, 31, 951–968.

Barker, K. 2008. Flexible boundaries in biosecurity: accommodating gorse in Aotearoa New Zealand. *Environment and Planning A*, 40, 1598–1614.

Bean, A. R. 2007. A new system for determining which plant species are indigenous in Australia. *Australian Systematic Botany*, 20, 1–43.

Cloke, P. and Jones, O. 2001. Dwelling, place, and landscape: an orchard in Somerset. *Environment and Planning A*, 33, 649–666.

Davis, M. A., Chew, M. K., Hobbs, R. J., Lugo, A. E., Ewel, J. J., Vermeij, G. J., *et al.* 2011. Don't judge species on their origins. *Nature*, 474, 153–154.

Downey, P. O. 2011. Changing of the guard: moving from a war on weeds to an outcome-orientated weed management system. *Plant Protection Quarterly*, 26, 86–91.

Gill, N., Klepeis, P., and Chisholm, L. 2010. Stewardship among lifestyle oriented rural landowners. *Journal of Environmental Planning and Management*, 53, 317–334.

Ginn, F. 2008. Extension, subversion, containment: eco-nationalism and (post)colonial nature in Aotearoa New Zealand. *Transactions of the Institute of British Geographers NS*, 33, 335–353.

Gould, S. J. 1997. An evolutionary perspective on strengths, fallacies, and confusions in the concept of native plants. In: Wolschke-Bulmahn, J. (ed.) *Nature and Ideology: Natural Garden Design in the Twentieth Century*, Washington, DC, Dumbarton Oaks Research Library and Collection, pp. 11–19.

Groves, R. H., Boden, R. and Lonsdale, W. M. 2005. *Jumping the Garden Fence: Invasive Garden Plants in Australia and their Environmental and Agricultural Impacts*, CSIRO Report prepared for WWF Australia.

Haraway, D. 2008. *When Species Meet*, Minneapolis, MN, University of Minnesota Press.

Harlan, J. R., and De Vet, J. M. J. 1975. Weeds and domesticates: evolution in the man-made habitat. *Economic Botany*, 29, 99–107.

Head, L. 2012. Decentring 1788: beyond biotic nativeness. *Geographical Research*, 50, 166–178.

Head, L., Atchison, J. and Gates, A. 2012. *Ingrained: A Human Bio-geography of Wheat*, Aldershot, Ashgate.

Head, L. and Muir, P. 2004. Nativeness, invasiveness and nation in Australian plants. *The Geographical Review*, 94, 199–217.

Head, L. and Muir, P. 2006. Suburban life and the boundaries of nature: resilience and rupture in Australian backyard gardens. *Transactions of the Institute of British Geographers NS*, 31, 505–524.

Head, L. and Muir, P. 2007. *Backyard: Nature and Culture in Suburban Australia*. Wollongong, University of Wollongong Press.

Hitchings, R. 2003. People, plants and performance: on actor network theory and the material pleasures of the private garden. *Social and Cultural Geography*, 4, 99–114.

Hitchings, R. 2006. Expertise and inability: cultured materials and the reason for some retreating lawns in London. *Journal of Material Culture*, 11, 364–381.

Hitchings, R. 2007. How awkward encounters could influence the future form of many gardens. *Transactions of the Institute of British Geographers NS*, 32, 363–376.

Hitchings, R. and Jones, V. 2004. Living with plants and the exploration of botanical encounter within human geographic research practice. *Ethics, Place & Environment*, 7, 3–18.

Hobbs, R. J., Arico, S., Aronson, J., Baron, J. S., Bridgewater, P., Cramer, A. A., *et al.* 2006. Novel ecosystems: theoretical and management aspects of the new ecological world order. *Global Ecology and Biogeography*, 15, 1–7.

Jones, O. and Cloke, P. 2002. *Tree Cultures: The Place of Trees and Trees in Their Place*, Oxford, Berg.

Larson B. 2008. Entangled biological, cultural and linguistic origins of the war on invasive species. In: Frank, R. M., Dirven, R., Ziemke, T. and Bernardez, E. (eds) *Cognitive Linguistics Research: Sociocultural Situatedness*, Berlin, Mouton de Gruyter, pp. 169–195.

Low, T. 2002. *The New Nature: Winners and Losers in Wild Australia*, Camberwell, Vic., Penguin.

NPWS (National Parks and Wildlife Service) New South Wales. 2002. *Urban Wildlife Renewal: Growing Conservation in Urban Communities*, Sydney, National Parks and Wildlife Service. Available at: www.nationalparks.nsw.gov.au.

Phillips, C. 2013. *Saving More Than Seeds: Practices and Politics of Seed Saving*, Aldershot, Ashgate.

Power, E. R. 2005. Human–nature relations in suburban gardens. *Australian Geographer*, 36, 39–53.

Thomson, E., Cleirigh, C., Head, L. and Muir, P. 2006. Gardeners' talk: a linguistic study of relationships between environmental attitudes, beliefs and practices. *Linguistics and the Human Sciences*, 2, 425–460.

Trigger, D. and Mulcock, J. 2005. Native vs exotic: cultural discourses about flora, fauna and belonging in Australia. In: Kungolos, A., Brebbia, C. and Beriatos, E. (eds) *Sustainable Planning and Development: The Sustainable World*, Vol. 6, Southampton, Wessex Institute of Technology Press, pp. 1301–1310.

Zagorski, T., Kirkpatrick, J. B. and Stratford, E. 2004. Gardens and the bush: gardeners' attitudes, garden types and invasives. *Australian Geographical Studies*, 42, 207–220.

7

EXPERIMENTS IN THE RANGELANDS

White bodies and native invaders

Cameron Muir

The 'scrub' country west of the Bogan River, in the rangelands of western New South Wales, was a problem for settlers, governments, and the whole civilising project right from the start. It was a place marked by massacre, environmental degradation, extinctions, and animal suffering. In the nineteenth century, great hordes of cattle, sheep and rabbits spilled into the interior of Australia and began eating up its grasslands. The loss of vegetation and the exposure and trampling of ancient soils, combined with drought, caused 'horrific' dust storms which left 4ft fences buried under sand (Griffiths, 2001). The stockmen and squatters accompanying the animals fought to the death with Aboriginal people for control of territory and scarce environmental resources. Killing was so common during the expansion of grazing on the plains that 'violence was normalised' (Broome, 2010: 46). Yet, for all this, the colonists did not recognise themselves and their agriculture as invaders.

Instead they built a society preoccupied with persuading itself and others of the legitimacy of its occupation of Australia. Whether something was native or not mattered less than whether it supported their vision for what the inland plains should be: in the late nineteenth, and for much of the early twentieth century, this was 'white' and agrarian. Introduced rabbits were despised as much as native dingoes. Aboriginal people, and Chinese settlers, in different ways, were both obstacles to a white civilisation for the antipodes. I will explore these ideas and obsessions in the story of two experiments carried out at the 'Bogan scrub' between 1896 and 1908.

In the first, 700 unemployed men were sent to the semi-arid woodlands along the Bogan River to clear 'invading' native vegetation so that others could settle there. It was a policy decision influenced by curiosity about the ability of white men's bodies to undertake manual labour in isolated and hot environments. In the second, the New South Wales Colonial Government established an experimental farm at Coolabah, amidst the saltbush plains and Bogan scrub (Figure 7.1), to investigate wheat-growing in dry country. One was corporeal with little planning or forethought. The other was cerebral and material and was carried out according to the formal principles of an emerging field of science. Both were about dealing with a 'hostile' environment and for an understanding of how Europeans could handle heat and aridity. The experiments arose from the intersection of local and empire-wide anxieties and were underpinned by provocative new ideas about the living world. How does scientific agriculture engage with the imperial and national agendas within which it operates? How do we see this manifest in place, and how does place itself shape scientific agriculture?

In September 1896, the New South Wales Labour Bureau sent 70 gangs of ten men each to the Bogan scrub near Nyngan (Anon, 1896c). Their task was to clear vegetation in preparation for the break-up of the large pastoral lease holdings and the settlement of small farmers. The Bogan River marks the transition between rich alluvial soils of the plains to its east, and older sandy red soils to its west. In some places the change is abrupt, but in other places there is a mix of black and red soils, shifting clays and sandy rises, and gilgais and cowals that hold water long

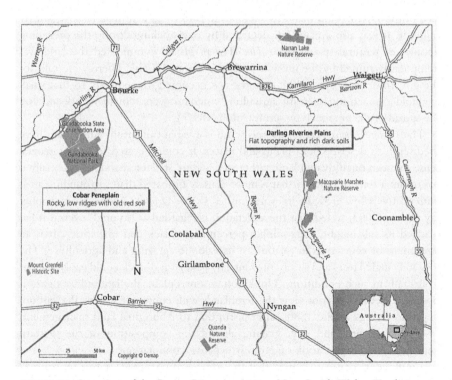

FIGURE 7.1 Location of the Bogan River in western New South Wales. To the west of the river are the semi-arid woodlands pejoratively named the 'Bogan scrub', to the east are the Darling Riverine Plains.

Source: Cartography by Damien Demaj.

after rain or a fresh flush in the river. It is a vast mosaic of grassed plains, soft earth, and thickets of white cypress pine, she oaks and dense shrubs. The red country supports semi-arid woodlands of mulga, poplar box, kurrajong, wilga, myall, and budda. Many of these native species are labelled 'woody weeds' today: undesirable plants that make grazing difficult. The red country at the Bogan represented the typical conditions with which settlers would have to contend in any plans for populating the dry interior: shallow sandy soils, hot summers, low variable rainfall, and little or no surface water. It had long been established there were no great inland rivers, and hopes that artesian water might 'redeem' the interior were fading. Would the continent's interior be an impediment to the colony's ambitions forever? Pastoral occupation had failed to create stable settlements. Agriculture might bring civilisation to the inland plains, but was it possible there? The New South Wales Government saw the Bogan scrublands as the place to find out.

In 1897, the New South Wales Under-Secretary for Lands, William Houston, journeyed to what were known as the 'West Bogan wastelands' to inspect the country. He reported that on either side of the route for 175 miles, 'it would be no exaggeration to say that the greater part of it is in its present condition a forbidding wilderness' (Anon, 1897b: 3). It was country that went unused in the face of the Divine Command to subdue the land or to legitimately possess it by tilling it. It was also a place left degraded by overstocking during the preceding decades of pastoral speculation. *The Western Herald* commented the land was 'temporarily ruined by the growth of scrub' (Anon, 1896b). There were calls for the government to ensure the country was occupied immediately so that small landholders practising scientific agriculture could carry out remedial work and stop the spread of the encroaching scrub (Anon, 1897b).

The *Oxford English Dictionary* says scrub is a variant of shrub. Australians have used scrub to describe both plants and places. It could be used in a very general sense to mean uncultivated land, either in a 'virgin' state or abandoned. Usually it has meant a tract of country that is more densely timbered than surrounding areas with relatively low growing trees and bushes. In *The Default Country*, lexicographer Jay Arthur (2003) noted that the vegetation is regarded as 'stunted' even if it has reached its full height, suggesting a perception of lack and inferiority. It is an assessment of vegetation in relation to its value for grazing and agriculture. The low and tangled trees and shrubs prevent easy passage and stock could easily wander from trails in such conditions. Unlike on an open plain, the landholder's gaze is obscured. Settlers did not view the scrublands with contempt at first. Perceptions began to change in the 1880s, after disruption to Aboriginal land management, restriction of fire, and over-grazing altered the composition of the western vegetation, and 'scrub' took on entirely negative connotations.

Landholders began to report that the scrub was changing. It was resisting all attempts to occupy it. The character of the semi-arid woodlands altered so that the hard, spiky and unpalatable plants dominated. One newspaper reporter following the Minister for Lands' visit to the western districts in the 1880s said of the dry woodlands at Byrock, north of Coolabah:

It is almost a treeless country . . . and absolutely level. The surface soil becomes pulverised by the traffic into an impalpable red dust, and the slightest agitation of the wind raises it in heavy clouds. It tinges everything around; in fact it is one of those places where every prospect displeases, and all but man is vile

(Anon, 1884: 14)

The scrub was increasing in density, it was closing over, and it was rejecting the European settlers. What is it about scrub that attracts suspicion and mystery? For Australian writer Ross Gibson the Brigalow scrub in Queensland is one immense crime scene, Australia's own badlands. It is a 'lair for evil' where 'malevolence flourishes naturally', or has been 'shoved in there since colonial times' (Gibson, 2002: 13). Scrub is unknown, unseen, unpopulated, and secret. The scrub is always silent. Historian Michael Cathcart provided a fascinating account of the symbolic importance of silence in *The Water Dreamers*, arguing, 'the silence expresses the moral condition of the land – the lethargy and stasis of a primordial forest. The silence is a dimension that exists before time' (Cathcart, 2009: 59). Only the arrival of whites could animate or bring time and history to the land. Scrublands received scorn from a society that was beginning to turn its gaze from the fertile and picturesque coast to the hostile and deficient interior. Perhaps it was predictable that the scrub itself became the problem, and not Australians' relationships with their environment. Was it inevitable that agriculture, as the new tool of social and environmental correction, would be called on to deal with the scrub problem?

Scrub-clearing was hard work. Men toiled in 40° heat using axes to fell and ringbark some of the hardest timber in Australia. Pastoral landholders would order their scrub-cutting gangs to cut a shallow ring around the outer bark, from as small as a single line that did not remove a chip, to about six inches wide. The trees lose the ability to send sugars to their roots and with this method take about three years to die. The landholders would leave the grey trees standing in the paddocks. Ringbarking for cultivation was even harder. Farmers or their workers would cut deep through the bark and sap-wood into the darker coloured heart-wood. If done this way, some species of the semi-arid woodland trees could die within a few months (Peacock, 1900b).

Chinese immigrants had provided labour for the pastoral industry from its earliest days and much ringbarking and suckering was carried out by Chinese scrubbing gangs (Campbell, 1923; Rolls, 1984; Bonyhady, 2002). The New South Wales Minister for Lands, Joseph Carruthers, was keen to see the Bogan scrub populated and agriculturally productive. He was also an advocate for federation of the colonies 'for the sake of white Australia' (Ward, 2006). Soon after the gold rushes of the 1850s, the New South Wales Colonial Government placed some restrictions on Chinese immigration. In 1896, the same year that the unemployed were sent out to the Bogan scrub, the New South Wales Government passed legislation which did not just restrict Chinese immigration, but banned all 'coloured' immigration. Without the option to exploit Asian and Islander workers, who would carry out the manual labour to prepare the ground for agriculture?

Would white people work for low pay, and perhaps more importantly, were they physiologically capable in the interior's heat and dust?

The white men doing the work of 'coloured' people soon found public sympathy in newspapers and in the labour movement. W. T. Goodge, a writer of light verse, wrote a poem about the plight of the 'struggle-for-lifers' for *The Bulletin* in 1899. These two verses illustrate some of the key concerns that emerged from the Bogan scrub experiment:

> Oh, come with me to the Bogan, boys,
> To the Bogan scrub so gay,
> Where our brethren toil on a hungry soil,
> At an Indian coolie's pay!
> . . .
> For the damper's tough in the Bogan, boys,
> And the beef's as hard as rocks,
> And the bull-dog ants get into your pants
> And eat your Sunday socks!
> No sinful pleasure is there, my lads,
> No wickedness there you know.

The scrublands have been a place of murder and massacre, a place where sin came easy, but where sin could be purged through battling the land, a place of perverse biological and corporeal reckoning, of environmental degradation and moral redemption. The scrub was no longer just a place in need of correction; it could also correct the defects of the unemployed through tough physical work. New ideas in biology created an obsession with the body and physical fitness. It is no surprise that 1896 was also the year of the revival of the Olympic Games. Towards the end of the nineteenth century physiological sciences such as anthropometry and phrenology could supposedly explain a range of physical and mental characteristics and health issues: 'symmetry of the body was thought to reflect physiological – even spiritual – fitness' (Park, 1994: 62). The men's labour was reshaping their bodies as well as reshaping their minds. All the while they were reshaping place.

In a survey of key texts on the history of masculinity and morals, J. A. Mangan and James Walvin argued that by the end of the nineteenth century ideas about 'manliness' had shifted from earlier ideals of the gentlemanly and Christian concerns of 'selflessness and integrity' to 'stoicism, hardiness and endurance' or 'neo-Spartan virility' (Mangan and Walvin, 1987: 1). In Australia, settlers felt the Aboriginal people and native animals were not worthy opponents. The 'kangaroos and other marsupials did not offer the classic challenge of the wild animals of Africa', wrote historian Libby Robin in *How a Continent Created a Nation* (2007: 42). The settlers killed both in great numbers but it did not bring honour. Robin argued that by the start of the twentieth century 'battling the land' stood in for the traditional challenge. Perhaps scrublands had taken on a symbolic, even hero-making, role.

By the end of the scheme the scrub-cutting gangs had cleared over 600,000 acres (Anon, 1899: 4). The Department of Lands had released 187,194 acres of the cleared land for Improvement Leases – the condition of the country was still so poor that those 200,000 acres could only support 49 mixed farms (Anon, 1898: 3). In 1900, of the half a million acres cleared, only 3,000 to 4,000 were under crop (Peacock, 1900b). The newspapers proclaimed the scheme a failure:

> Persistent attempts have been made to make it wealth-producing, but in a large number of cases the experiment has been heart-breaking. Hundreds of thousands of pounds have been lost there, and many a man discouraged almost beyond redemption
>
> *(Anon, 1900a: 4)*

Some of the scrub cutters were directed to clear the major stock routes in the west as well. Dubbo's *Daily Liberal* complained, 'the result now appears in the steady and indiscriminate destruction of almost all the shade trees along the routes' (Anon, 1900b). Some residents took 'grim consolation' in the fact that the ring-barking work was so poorly done it appeared not every tree was going to die (Figure 7.2).

In 1896, Australia led the world in formalising the idea of whiteness. In the words of Marilyn Lake and Henry Reynolds, Australia embarked on a 'radical new departure in international relations' when the New South Wales Colonial

FIGURE 7.2 A photo of semi-arid woodlands, or 'scrub country', near Coolabah, taken in winter 2013.

Source: Cameron Muir.

Government introduced the *Coloured Races Restriction and Regulation Act* in 1896 (Lake and Reynolds, 2008: 144). The other colonies soon followed suit, and at Federation the new national government passed the *Immigration Restriction Act 1901*. These Acts had their origins in anti-Chinese protests and violence from the 1850s, when miners objected to competition from these 'industrious' immigrants (Department of Immigration and Citizenship, 2007). The New South Wales and Victorian Governments introduced some restrictions on Chinese immigration soon after, as did other countries around the world, including the United States and Canada.

What was radical about the 1896 legislation in New South Wales was that it banned 'all persons belonging to any coloured race inhabiting the Continent of Asia or the Continent of Africa, or any island adjacent thereto, or any island in the Pacific Ocean or Indian Ocean', irrespective of a country's power, or status as an ally, or even if an individual was a British subject (Lake and Reynolds, 2008: 144). This angered Japan and India, and embarrassed Britain (Morris-Suzuki, 2008). It formally divided the world into white and not white for the first time (Lake and Reynolds, 2008). Fears fuelled by economic self-interest made the demands for restrictions urgent, but even these were influenced by ideas about race.

It is hard to appreciate how profound a rupture the beginning of the biological revolution must have been for nineteenth-century society. As Europeans made their first incursions into Australia's interior, geologists began to understand that the Earth must be millions of years old, not a few thousand. In *Hunters and Collectors*, historian Tom Griffiths conveyed their dramatic expansion in perspective: 'in the two hundred years following the European invasion of Australia, the known age of the earth increased from about 6000 years to 4.6 billion' (1996: 8). Soon after the geologists made their revelations, the naturalists put forward persuasive arguments for an even more radical theory – humans were just another animal and differentiation of species was a random, contingent process. 'Nineteenth-century evolutionists', wrote historian Edward J. Larson, 'envisioned the earth as a grand laboratory or workshop of organic development: a shimmering sphere of life spinning in a vast universe' (2004: xiii). It was the most fundamental shift to how we viewed ourselves and other living creatures, we 'became interconnected competitors rather than separate creations' (ibid.). Biological thought has flourished over the last 150 years. Larson wrote: 'We now live in the shadow – or the illumination – of this modern biologic worldview' (ibid.).

Where once God had made the world and all its creatures final and stable, Cuvier demonstrated a world in which species could become extinct. Lamarck argued that species change in response to environment, and Darwin recognised this process as historical. Niles Eldredge of the American Museum of Natural History wrote that Darwin transformed 'the prevailing view of stability – of the earth, of all the species on earth, and not least the stability of society's strata – into a picture of motion' (2006: 8). The hierarchy of species was a result of a dynamic process in time – the 'lower' species were older, the 'higher' newer, and 'the struggle between them created ever "higher" forms' (Lindqvist, 1996: 120). Different 'races' of humans could be ranked separately in the list of species. Some might be an older 'type' or form and therefore the link between advanced humans and other animals, or some groups could have degenerated from the higher type. It was a classification system that reflected existing power structures and cultural perceptions of superiority (Eldredge, 2006).

Now that human difference had a biological basis, it meant that 'white races' must have distinct biological characteristics. In *The Cultivation of Whiteness*, Warwick Anderson argued that previously 'white' in Australia had meant British ancestry, but from the 1880s 'whiteness' became a 'type, mobile and standardized' (2003: 2). Skin colour was too unreliable a determinant, it had to be something more, but it was elusive. Perhaps it was 'a typical bodily constitution or temperament', a thought-style, head circumference, predisposition or resistance to diseases (ibid.). The biological basis for whiteness also encompassed its vulnerabilities and limitations, not just what made it different and superior. Foreign environments could affect white bodies. A common belief was that white people were unable to live and work successfully in the humid and disease-ridden tropics, that there was 'something in the tropical climate inimical to Europeans' (Smith, 1899). The effects of the environment of the dry interior were more ambiguous. It was too hot to be temperate, but it was not humid enough to be tropical. The little rain that fell evaporated rapidly. When scores of inland settlers died in the heat in 1896, Bourke's *Western Herald* proposed the solution was a simple matter of cultural adaptation to the environmental conditions. It implored people to shed the traditions of the Old Country, such as hot meals, heavy clothes and business hours during the middle of the day when it was hottest, to mitigate against 'the terrors of the climate' (Anon, 1896a: 2). The editorial argued they were not bound by biological constraints; instead, how they chose to live with the heat was within their control.

When immigration restrictions were enacted in the colonies and then in the new nation, to protect Australia, 'for all time from the contaminating and degrading influence of inferior races' (Isaac Isaacs, cited in McGregor, 2008: 5), it had implications for agriculture in Australia. Queensland asked for a Royal Commission into the circumstances of tropical agriculture but this was denied. In 1906 and 1907 nearly 4,000 islanders were deported (Willard, 1967). Agriculture was intricately involved in the production and protection of 'whiteness' immediately after the policy changes, and for the next few decades. The deployment of unskilled labour to the Bogan scrub and the experiment farm at Coolabah were the first engagements with the practicalities of populating and protecting a white Australia.

In 1898, Robert Peacock set out for the new experiment farm near Coolabah. As manager of the farm, Peacock was charged with carrying out no small task. His mission was to gain 'the knowledge which it will be necessary to possess before the millions of acres of now almost useless land, of which the farm is typical, can be turned to good account' (Mylrea, 1990). Fifteen hundred acres were presumed for the purpose, and W. S. Campbell of the Department of Agriculture chose 200 acres on which to make a start. The farm was to give 'special attention' to wheat cultivation (Anon, 1897a: 5). In the late nineteenth century, plant-breeders crossed wheat varieties to suit the conditions of the New World lands. They made vast tracts of 'virgin' territory available for cultivation of European plants, consolidating the 'Great Grain Invasion' which saw the harvests of those plants shipped back to Europe, and which lowered the price of grain foods and devastated the farmers of Europe (Olmstead and Rhode, 2006). Growing wheat at the Coolabah experiment farm would use place in two ways. It would allow the plant-breeders to select seeds based on which plants survived the extreme conditions, and they hoped the plants would adapt to the conditions through their interaction with the environment.

Peacock would need to collaborate with the Department's specialist staff to make it possible to grow wheat in an environment where 'the absence of rain' had made it impossible (Anon, 1897a). Coolabah at that time was 150 miles beyond the limits of the wheat-belt. Rainfall in one year barely totalled 80 millimetres. If crops grew here, they would grow anywhere. The Minister instructed the American-born plant pathologist Dr Nathan Cobb to 'undertake exhaustive experiments with a view of selecting a wheat that will adapt itself to the climate, and be sufficiently early to benefit by the rainfall' (ibid.). The independent wheat-breeder William Farrer, by then on the Department's books, began wheat trials using the conditions at the Coolabah farm to produce drought-tolerant wheat varieties.

The Department's timing for the establishment of the Coolabah Experiment Farm was unfortunate. Experimentalists began cultivating their wheat plots just as eastern Australia entered what became known as the Federation Drought. Farrer was frustrated with the low yields of the 'Macaroni', or durum wheat, he had selected for the dry and hot conditions in the interior. He tried North African, Mediterranean, West Asian and Central European wheats with little success (Peacock, 1904a). The wheat trials did not fare any better when the rains returned. Typical of the extreme variability of the western country, when the drought broke, it was with fierce storms and flooding. The farm's manager, Peacock, reported that in September 1903 heavy rain caused a nearby cowal to overflow and flood some of the trial plots, and strong winds bent many of the wheat plants. The next month hail storms tore up the plants, and with the unusual amount of rain, rust set in. The rain turned the broken and bare ground hard. With so much rain after such a long drought the grasses and shrubs boomed, but so did the wildlife, including mouse populations. The mice climbed the wheat stalks and attacked the grain, they used the string bands of the sheafs to make their nests, and 'after wheats were sown they would follow the drills, and scratch and devour the grain' (ibid.: 1120). When the remaining seeds germinated, they ate the shoots off to the level of the ground. They also destroyed the haystacks, ate the insides out of the melons and pumpkins, and polished off a crop of cowpeas before they ripened. The mice 'appeared to be everywhere, in houses, barns, stables and fields, and would eat almost anything' (ibid.). Peacock's reports were rarely positive.

Robert Peacock knew dry country. He was a self-taught generalist raised on a mixed farm near Bathurst. He knew the soils in the west ranged between poor red sandy loams and richer but difficult to cultivate black cracking soils. He knew grasses could dry out in a drought and get carried away by the wind. He knew rainfall was uncertain, and conditions 'decidedly arid' (Peacock, 1900a). As manager of the Coolabah Experiment Farm, he did little wheat-breeding research. The country there forced a change in the focus of his work at the farm. He was shocked at the state of the western country after years of pastoral occupation. The varied and 'luxuriant' grasses of the semi-arid plains and woodlands, such as Kangaroo in the red country and Mitchell on the black soils, had been over-grazed and were confined to a few protected areas and low-lying gilgais and cowals. The edible shrubs that provided extra fodder in drought had also been eaten bare and trees had been ringbarked. Pine scrub, bimble box seedlings, budda and acacias had 'taken complete possession to the thorough exclusion of even the worst grasses' (ibid.: 655). On the river plains, on which woody scrub does not grow, saltbushes began to be replaced by a related but spiny and unpalatable shrub called Roly-Poly. Many of the 'once coveted western properties' had been abandoned, and millions of acres were now 'beyond the scope of profitable occupation' (ibid.: 652).

There he was, 31 years old, alone in the far west, in charge of a new experiment farm and assigned with the remarkable responsibility of undertaking what was perhaps the most ambitious project scientific agriculture had attempted yet. He stood amongst the forlorn remnants of former abundance, on red ground strewn with parched fragments of soil-holding bushes, surrounded by grey and black dead timber, and pulverised grasses. Tumbleweeds rode the wind from out of the flats and collected on fences or in the thickening scrub. Most worrying of all, the 'perpetual summer winds' were creating 'ever-increasing' scalded plains of bared subsoil. This country called on him to talk straight. Writing in the *Agricultural Gazette of New South Wales* in 1900, Peacock declared the pastoral occupation of the west was 'a period of deterioration unexampled in the history of New South Wales' (1900a: 652). The numerous and visible examples of environmental degradation were 'too familiar landmarks, resulting from the mistakes of the past, and calculated to teach valuable lessons to those willing to listen to the voice and teachings of Nature'. In Peacock's view, nature was an active presence. He gave it a voice. These were the admonitions of a professional in a tough situation, invoking the higher authority of 'Nature' against those whom he blamed for recklessly exploiting the drylands.

The isolated experimentalist's words caught the attention of government officials and in 1901 Peacock was called on as a witness in a Royal Commission investigating the failure of settlement in the Western Division. In Peacock's evidence, the Commission heard explanations for the changes in the western environment that were based on scientific theory. Rather than listing 'non-edible scrub' as another hardship forced on settlers by a deficient and backward Australian environment, Peacock explained, 'Nature, in order to recover her equilibrium, is at present producing vegetation capable of adapting itself to its natural surroundings by its unpalatableness and protective spiny growths' (ibid.: 654). Peacock was a Darwinist. Plant life was evolving to survive the pressures landholders were placing it under.

Peacock had great respect for the native vegetation of the semi-arid interior. It was uniquely adapted to its environment, it could teach settlers about the long-term climate, it was drought-resistant, the number of species was 'truly remarkable', and 'in its primeval condition was admirably adapted for the support of animal life throughout prolonged periods of dry weather' (ibid.: 652). His *Agricultural Gazette* article seethed with an undercurrent of outrage that exploitative stocking methods had left 'many varieties . . . well-nigh extinct'. In tandem with wheat experiments, he began planting trial plots and paddocks with various saltbushes.

The genus name for saltbushes is *Atriplex* and different species grow around the world in semi-arid and arid climates. They tolerate soils with high salt content and their leaves are salty to taste. They have been consumed in human diets for millennia but recently they have been valued more for their role in renewing degraded and salt-affected soils in drylands. In 1904, Peacock had planted 60 acres of saltbushes at the Coolabah Experiment Farm (Figure 7.3). The most valuable for pastoral properties, and probably the most well-known species, was Old Man Saltbush (*Atriplex nummularia*). Peacock's Old Men grew 3 metres tall and stayed green for the duration of the drought. Protecting the plants and allowing them to reach that height meant the higher branches were beyond the reach of the sheep and this prevented over-grazing. It provided shelter for animals, produced an abundance of seed, and was easy to cultivate.

FIGURE 7.3 Saltbush cultivated under the management of Robert Peacock in 1900. The original caption in his report read, 'Saltbush that is given a chance in the Coolabah Experiment Farm Paddock.'

Source: *Agricultural Gazette of New South Wales*, National Library of Australia.

In droughts, saltbushes survived long after the grasses had parched and died. Successful occupation of the interior depended on the providence of the native vegetation that had adapted to the environment. Peacock implored, 'the West must look after its valuable plants, and not ruthlessly destroy those which make it valuable' (1904b: 220). Many species did not return after the drought. They had been over-grazed and some became locally extinct. Peacock wrote that the people of the interior should 'make good in some small measure the mistakes of the past by the cultivation and conservation' of the saltbushes (ibid.).

The international exchange of plants among agricultural institutions in the late nineteenth and early twentieth centuries is usually associated with wheat. Australian saltbushes, however, found favour among agricultural botanists and experimentalists who began to regard the plants as useful fodder for stock in semi-arid and arid regions, especially in the United States and South Africa (Frawley and Goodall, 2013). After his years of experiments, Peacock was convinced that 'the Australian saltbushes have no rivals in drought-resistance' in the world (Peacock, 1904b). He sent seeds of seven different saltbushes to the United States Department of Agriculture, as well as two species of Mitchell Grass (United States Department of Agriculture, 1902). These were distributed mainly in California and Arizona as a food source in dry areas, and to help stabilise sandy or alkaline soils. The saltbushes still grow in isolated areas in these states, along roadsides, on sea bluffs, and coastal plains and basins. When Los Angeles airport was expanded and new jets roared above the suburbs surrounding it in the 1960s and 1970s, the beachside settlement of Palisades del Rey and its 822 homes were abandoned. Old Man Saltbush found a refuge in the vacant lots and overgrown streets, along with an endangered butterfly species, El Segundo Blue. These are the material remains of international cooperation, the sharing of scientific knowledge and genetic material, and the reckoning of places.

In 1908, an article in *The Sydney Morning Herald* stated that when anyone thought of the plains between Nyngan and Bourke, the immediate vision was of 'big stretches of barren land' and the experience of failure, and prompted the people of Australia to ask 'whether it is all worthwhile' (Anon, 1908a). The author observed that the crops in Dubbo and Wellington, in the recognised wheat-belt, were 'parched plains and scaled patches'. One hundred and fifty miles past that, 'out towards that wilderness of sand and all that is desolate; the sepulchre of so many hopes, ambitions, and gallant endeavours' was the Coolabah Experiment Farm. For years, Peacock had described the dismal state of the wheat-growing experiments in his annual reports to the New South Wales Department of Agriculture. In 1902, he recommended that 'wheat-growing in large areas should be discontinued here' (Peacock, 1902). By 1906, even the wheat experimentalist George Sutton agreed: the best that could be hoped for the Bogan scrub country was to grow cereals for hay to provide feed in drought. Peacock wanted to continue the saltbush trials and resisted the closure of the farm, arguing the experiments were incomplete and they would lose the opportunity for long-term comparisons (Anon, 1908b). Saltbush cultivation for grazing was not a priority for the Department. In 1908, at the time the closure was confirmed, there were still 'millions of acres yet untouched in this state which must be broken by the wheat-farmer's plough' (Anon, 1908a). Sutton was determined to continue the wheat experiments in an effort to extend the wheat-belt. The Coolabah farm closed and the wheat experiments withdrew to a new experiment farm at Nyngan.

The country at Coolabah forced a new imperative. Instead of formulating a system in which to carry out the transformation of idle scrublands into orderly agricultural fields, Peacock felt his first priority was to 'reclaim these wastelands and, if possible, bring them to a semblance of their former condition' (Peacock, 1900a: 652). His task became one of rescue, not revolution. This is the pattern Australian science for agriculture followed in its early days. The small Department of Agriculture was pulled between the demands of popular development rhetoric and the need to respond to the on-the-ground circumstances of settler farming and Australian environmental conditions.

At the end of the nineteenth century the government was concerned that the resources it held were being depleted, primarily in pastoral operations, but also in the way burgeoning crop farming was being practised. The main attraction of scientific agriculture was its promise of allaying some of these environmental problems, along with its potential civilising influence on the culture of inland settlement. Some public figures, however, perhaps influenced by the popular press, and perhaps speaking to popular sentiment, persisted with dreams of whole-scale and widespread transformation of the dry interior. This kind of talk was alluring and infectious, and often the Department positioned itself as the institution with the knowledge and expertise to make such projects a reality. Their funding depended on the results being relevant to policy. Generally, however, the early experimentalist staff were practical in their work and its scope.

Despite the public anxiety to settle for cultural and political reasons, many of the experimentalists and other Department staff were sensitive to place. They were critical of environmental over-exploitation and farming practices that tried to profit only in the short term. In *The Delicate and Noxious Scrub*, retired CSIRO scientist Jim Noble has argued that Peacock's work 'probably represent[ed] the first formal rangelands research undertaken in Australia' (1997: 36). Peacock was not alone in seeing the agricultural project as a restorative one. In response to the drought and harvest failure in the Russian Steppes in 1891, agricultural scientist Vasilii Dokuchaev began researching the effects of human exploitation of the plains. He was critical of exploitative farming practices and devised plans to repair the environmental degradation of the Steppes (Moon, 2005). In the United States, agricultural reformers such as Theodore Roosevelt strived to establish a conservation ethic that treated 'wise-use' and aesthetic preservation as belonging to the same long-term civilising project (Tyrrell, 2012/2013).

Landholders continued to clear and over-stock semi-arid woodlands over the course of the twentieth century. In 1938, biologist Francis Ratcliffe published his classic *Flying Fox and Drifting Sand*, based on his travels across the rangelands. He criticised graziers for over-stocking and the destruction of native vegetation it caused. By the 1960s and 1970s landholders were using heavy chains strung between two tractors to carry out broad-scale clearing. They also tried aerial and ground spraying with 2,4,5-T and 2,4-D, the main ingredients of Agent Orange. In the 1980s, rangeland scientists described the semi-arid woodlands as a 'resource under siege' (Joss *et al.*, 1986). In 1997, the New South Wales Government introduced restrictions on the clearing of native vegetation. Landholders were outraged because they believed the growth of woody native vegetation was unnatural. In their view, these native species were invaders, devaluing grazing properties. Landholders at the Bogan formed the Rural Community Survival Group and locked out government environmental compliance managers from their properties. To them, illegal clearing was not degradation, it was restoration.

A different conservation ethic continues as well. In October 2010, I went to a property near Girilambone and Coolabah belonging to grazier, Grant MacAlpine. Ray Thompson of the Central West Catchment Management Authority demonstrated how he uses a laser level to work out the gradients for doing 'water-ponding' works. The oval-shaped banks of earth retain moisture and catch the seeds of native grasses and shrubs. When Ray was finished doing the mapping Grant started up his massive grader and showed how he creates banks for water-ponding by turning over the earth. He spends hours collecting saltbush seeds from his existing plants and distributes the seed over the turned earth.

In the early twentieth century, Robert Peacock, the manager of the Coolabah experiment farm, was adamant that the Department of Agriculture and landholders should focus on regenerating the native vegetation. Over one hundred years later, that's what Ray and Grant were doing (Figure 7.4).

FIGURE 7.4 Catchment Officer Ray Thompson standing next to saltbush he cultivated on private property in partnership with local landholders.

Source: Cameron Muir.

References

Anderson, W. 2003. *The Cultivation of Whiteness: Science, Health and Racial Destiny in Australia*, New York, Basic Books.
Anon. 1884. Ministerial visit to the northwest. *Sydney Morning Herald*, 6 December, p. 14.
Anon. 1896a. The heat wave. *The Dubbo Liberal and Macquarie Advocate*, 15 January, p. 2.
Anon. 1896b. Scrub lands and the unemployed. *Western Herald and Darling River Advocate*, 22 August.
Anon. 1896c. Work for the unemployed. *Brisbane Courier*, 1 September.
Anon. 1897a. Experimental farm at the Bogan Scrub, *Sydney Morning Herald*, 16 December, p. 5.
Anon. 1897b. The Bogan Scrub. *Sydney Morning Herald*, 4 October, p. 3.
Anon. 1898. West Bogan Scrub country. *Sydney Morning Herald*, 5 August, p. 3.
Anon. 1899. New South Wales unemployed. *Brisbane Courier*, 11 November, p. 4.
Anon. 1900a. The problem of the West. *Clarence and Richmond Examiner*, 7 August, p. 4.
Anon. 1900b. The 'unemployed' and the tree-ringing, *Daily Liberal*, 19 September, p. 2.
Anon. 1908a. Coolabah State Farm: its lessons. *Sydney Morning Herald*, 8 July.
Anon. 1908b. On the land. *Sydney Morning Herald*, 11 August.
Arthur, J. M. 2003. *The Default Country: A Lexical Cartography of Twentieth Century Australia*, Sydney, UNSW Press.
Bonyhady, T. 2002. *The Colonial Earth*, Carlton, Vic., Melbourne University Press.
Broome, R. 2010. *Aboriginal Australians: A History Since 1788*, Crows Nest, NSW, Allen & Unwin.
Campbell, P. C. 1923. *Chinese Coolie Emigration to Countries Within the British Empire*, London, King.
Cathcart, M. 2009. *The Water Dreamers: The Remarkable History of Our Dry Continent*, Melbourne, Text Publishing Company.
Department of Immigration and Citizenship.2007. *Fact Sheet 8: Abolition of the 'White Australia' Policy*, Canberra, Department of Immigration and Citizenship. Available at: http://www.immi.gov.au/media/fact-sheets/08abolition.htm (accessed 25 August 2009).
Eldredge, N. 2006. *Darwin: Discovering the Tree of Life*, New York, W. W. Norton & Co.
Frawley, J. and Goodall, H. 2013. Transforming saltbush: science, mobility and metaphor in the remaking of intercolonial worlds. *Conservation and Society*, 11, 176–186.
Gibson, R. 2002. *Seven Versions of an Australian Badland*, St. Lucia, Qld., University of Queensland Press.
Griffiths, T. 1996. *Hunters and Collectors: The Antiquarian Imagination in Australia*, Cambridge, Cambridge University Press.
Griffiths, T. 2001. One hundred years of environmental crisis. *The Rangeland Journal*, 23, 5–14.
Joss, P. J., Lynch, P. W. and Williams, O. B. (eds) 1986. *Rangelands, a Resource Under Siege: Proceedings of the Second International Rangeland Congress*, Canberra, Australian Academy of Science.
Lake, M. and Reynolds, H. 2008. *Drawing the Global Colour Line: White Men's Countries and the International Challenge of Racial Equality*, Cambridge, Cambridge University Press.
Larson, E. J. 2004. *Evolution: The Remarkable History of a Scientific Theory*, New York, Modern Library.
Lindqvist, S. 1996. *Exterminate All the Brutes*, New York, New Press.
Mangan, J. A. and Walvin, J. (eds) 1987. *Manliness and Morality: Middle-Class Masculinity in Britain and America, 1800–1940*, Manchester, Manchester University Press.
McGregor, R. 2008. *The White Man in the Tropics*, Thuringowa. Available at: http://www.townsville.qld.gov.au/resources/3873.pdf (accessed 3 September 2009).

Moon, D. 2005. The environmental history of the Russian Steppes: Vasilii Dokuchaev and the harvest failure of 1891. *Transactions of the Royal Historical Society*, 15, 149–174.

Morris-Suzuki, T. 2008. Migrants, subjects, citizens: comparative perspectives on nationality in the prewar Japanese Empire. *The Asia-Pacific Journal: Japan Focus*. Available at: http://www.japanfocus.org/-Tessa-Morris_Suzuki/2862.

Mylrea, P. J. 1990. *In the Service of Agriculture: A Centennial History of the New South Wales Department of Agriculture 1890–1990*, Sydney, NSW Agriculture & Fisheries.

Noble, J. C. 1997. *The Delicate and Noxious Scrub*, Lyneham, A.C.T., CSIRO Wildlife and Ecology.

Olmstead, A. L. and Rhode, P. W. 2006. Biological globalization: the other grain invasion. *ICER Working Papers*, ICER, International Centre for Economic Research.

Park, R. J. 1994. A decade of the body: researching and writing about the history of health, fitness, exercise and sport, 1983–1993. *Journal of Sport History*, 21, 59–82.

Peacock, R. W. 1900a. Our western lands: their deterioration and possible improvement. *Agricultural Gazette of New South Wales*, 11, 652–657.

Peacock, R. W. 1900b. Western agriculture. *Dubbo Liberal*, 17 January.

Peacock, R. W. 1902. Report of the Manager, Coolabah Experimental Farm. *Agricultural Gazette of New South Wales*, 13, 575–577.

Peacock, R. W. 1904a. Report of the Manager, Coolabah Experimental Farm. *Agricultural Gazette of New South Wales*, 15, 1118–1121.

Peacock, R. W. 1904b. Saltbushes, their conservation and cultivation. *Agricultural Gazette of New South Wales*, 15, 211–220.

Robin, L. 2007. *How a Continent Created a Nation*, Sydney, UNSW Press.

Rolls, E. C. 1984. *A Million Wild Acres*, Ringwood, Vic., Penguin.

Smith, C. H. 1899. White men in the tropics. *Independent*, 9 March.

Tyrrell, I. 2012/2013. To the halls of Europe: Theodore Roosevelt's African jaunt and the campaign to save nature by killing it. *Australasian Journal of Ecocriticism and Cultural Ecology*, 2.

United States Department of Agriculture. 1902. Seeds and plants imported through the section of seed and plant introduction for distribution in cooperation with the agricultural experiment stations. *Bureau of Plant Industry Bulletin*, 5.

Ward, J. M. 2006. Carruthers, Sir Joseph Hector McNeil (1856–1932). *Australian Dictionary of Biography, Online Edition*. Available at: http://adbonline.anu.edu.au/biogs/A070582b.htm (accessed 6 August 2009).

Willard, M. 1967. *History of the White Australia Policy to 1920*, London, Cass.

8

THORNY PROBLEMS

Industrial pastoralism and managing 'country' in northwest Queensland

Haripriya Rangan, Anna Wilson and Christian Kull

Introduction

The landscapes of northwest Queensland have seen over a century of change induced through the creation and establishment of a pastoral industry. New plants, animals, people, infrastructure, technologies, and ideas of landscape have reshaped both the land and industry at the centre of the region's livelihoods. The pastoral industry in northwest Queensland developed over the latter half of the nineteenth century, an era of settlement expansion in the 'northern' frontier and a time when ideas of 'acclimatisation' were being embraced by Australian colonies and settlers alike. Three prickle bushes were introduced to northwest Queensland during this epic period to assist the process of pastoral expansion and acclimatisation: prickly acacia (*Acacia nilotica*), Parkinsonia (*Parkinsonia aculeata*), and mesquite (*Prosopis spp.*). At the time of introduction, these plants were seen as resuscitating nature and improving the country's natural resources for supporting the sheep industry. The industrial pastoralism of today centres on cattle and these plants are no longer considered saviours but are regarded as invasive species harming the land and native biodiversity.

This chapter explores the shift in attitude towards introduced prickle bushes in relation to changes in pastoralism and ideas of land management in northwest Queensland. We argue that the current preoccupation with clearing 'invasive' prickle bushes is not so much about the plants themselves, but rather the manifestation of anxieties about the sustainability of pastoralism in this region. These anxieties are expressed in terms of the relative virtues of 'native' and 'alien' plants and how they affect the landscape for raising cattle and sustaining the pastoral economy. Changes in government discourse regarding land management as 'land care' and 'caring for our country' have provoked pastoralists in this region to debate the ways of managing 'country' for cattle in the present and future. The following sections trace the history of changes in values regarding introduced plant species in Australia and the economic shifts that have transformed pastoralism in northwest Queensland. We contextualise the emergence of the prickle bush 'problem' alongside changing discourses of land management and pastoralist experiences of tackling prickly trees on their land.

Native and alien species in Australia

Most European settler societies have had a long and awkward relationship with concepts of 'native' and 'nature'. Both concepts have posed existential challenges to ideas of authenticity and being in the land. Early Anglo-Celtic settlers in Australia described their encounters with natives and nature in terms of their strangeness and lack of culture. Marcus Clarke, a renowned nineteenth-century Australian journalist and novelist, wrote, 'In Australia alone is to be found the Grotesque, the Weird, the strange scribblings of Nature learning how to write' (Clarke, 1909: iii). The strange scribblings of nature were in the form of peculiar animals and plants. Joseph Banks described kangaroos as, 'almost the Size of a middling Sheep, but very swift and difficult to catch' (Banks, 1779, cited in Turner, 1968: 5). In his journals of exploration into Queensland, Sir Thomas Livingstone Mitchell, wrote of 'Various very remarkable shrubs new and strange to me' and of 'trees of a very droll form' that 'looked very odd' (Mitchell, 1848: 219). Yet at the same time, settler desire to rewrite the scribblings of nature into a narrative of potential profit was overwhelming. As James Cook surveyed the prospect of settlement in this new and strange land, he reassured his audience in England that,

> it can never be doubted but what most sorts of Grain, Fruits, Roots &c of every kind would flourish here were they brought hither, planted and cultivated by the hand of Industry and here are Provender for more Cattle at all seasons of the year than can be brought into this country.
>
> *(Cook 1770, in Turner, 1968: 4–5)*

Nineteenth-century settlement in Australia grappled with encounters of strange native-ness, desire for profit, and yearning for familiar alien-ness through pastoralism and acclimatisation. Acclimatisation was the way of 'improving' the land by collecting plants 'from every quarter what is adapted to our soils and climate to new clothe our adopted country' (Barron Field, 1822, cited in Anderson, 2003: 430). The Queensland Acclimatisation Society, established in 1862, urged the colonial government to support the introduction of new animals and plants for settlers to realise the profit and potential from the land. It declared:

> The importation and acclimatisation of new animals and vegetable productions of a useful description as necessary to the prosperity of the country [are] obvious, from the fact that the chief producing interest in the colony owes its origins to the importation of animals not indigenous to Australia, and

that almost all future products – such as sugar, coffee, tobacco, cotton &c., whether for home consumption or exportation, will have been similarly originated.

(Anon, 1866: 5)

Pastoralism and acclimatisation became the narratives through which Australian colonies attempted to rewrite nature and write out natives from the landscape. They provided the means for 'civilising' the landscape with sheep, cattle, and exotic plants and thus securing the land from native incursions of all kinds. The industries arising from improvement and acclimatisation projects formed the heart of the Australian settler economy and identity for over a century. Despite the majority of the population being concentrated in coastal cities and tied to the global imperial economy, Australia's cultural identity was seen as being forged by the grit and ingenuity of pastoralists who wrenched success from battling against the alien and hostile native nature of the land (Bean, 1910; Tyrrell, 1999).

The metaphor of 'battle' carries an enormous amount of symbolism in Australian national consciousness both in terms of its Anglo-Irish settler history and everyday life on the land (Powell, 1988). Robin (2007: 6) notes that the popular celebration of Australian identity centres on being 'perversely proud of "battling the elements" and making life in a contrary environment', rather than 'coming to terms with Australian nature with all its richness, and its limitations and exceptionalism' (ibid.: 205). Settlers who battled against the most 'Australian' aspects of nature in the 'Outback' – the searing heat, droughts, floods, dingoes, natives – became the quintessential icons of Australianness. Paul McGuire described the drive to battle against the land as an 'Australian Thing', writing:

> She was hostile to me. She has been subdued to our human purposes only with fierce struggle, and in that struggle has appeared the Australian Thing. A culture is something more than the flower of civilisation. It is a whole way of life. It shapes men's characters and it shapes their landscapes: it is itself shaped by all the conditions of time and place and by men's intellectual and moral energies and values . . .
>
> *(1939: xvii)*

The relationship between Australianness, natives, and native nature underwent a radical shift from the 1960s onwards with the political struggles for Aboriginal rights, the rise of environmentalism, and the reimagining of national identity. As the White Settler nationalism began to be revised into a more inclusive narrative of the Australian nation, the focus of battle swung away from natives and native nature towards tackling 'alien' nature. The new imagining of Australian nationalism enthusiastically embraced native nature and turned against introduced plants and animals that appeared to threaten the country's unique environment. Despite the pastoral and agricultural economy being centred on introduced crops and animals, national economic development emphasised the importance of quarantine in keeping out non-native species that could potentially carry exotic diseases and organisms (Muller *et al.*, 2009). The rise of the global discourse of biodiversity conservation during the 1980s was embraced by environmentalists and agricultural policy-makers alike through the metaphor of invasions by alien species. The first newsletter of the IUCN's Invasive Species Specialist Group (ISSG) entitled 'Aliens' declared that 'the spread of non-indigenous plant species' was 'among the greatest threats to Australia's biodiversity . . . Almost every major ecosystem has been extensively altered . . . and degradation continues as these species infill and expand their range' (ISSG, 1995: 13).

Although contemporary imaginings of Australian national identity celebrate Indigenous and multiple cultures of immigrants, the legacy of the awkward relationship between settler and native nature persists in new ways of engaging with the land. Native flora and fauna have gone from being grotesque and inadequate to beautiful and cherished for their intrinsic native-ness, but their value continues to be measured in economic terms in relation to agricultural and pastoral industries. Protecting Australian nature continues, at heart, to be about defending a particular culture of pastoralism and agriculture on the land as symbolic of the country's national identity (Davison and Brodie, 2005).

National land narratives: from improvement to care

The terms 'land' and 'country' are often used interchangeably in everyday conversation, but the words convey very different sensibilities of relationships between people and place. Broadly speaking, land conveys a sense of ownership, the possibility of delimiting territory, establishing a claim to it, making it a commodity. Country, on the other hand, carries a meaning that is much more than possession. It places individuals in relation to a living world encompassing human and non-human species; it is constantly produced and sustained through these living relationships. It is interesting that Australian Aboriginal English chooses 'country' as the word to refer to areas that would, in mainstream English, be commonly considered land. 'Country', not preceded by any article indicating possession, is the Aboriginal English expression of 'a living world that has its sources in the past and in place, that is in the process of becoming, that works towards interspecies relationships that are "always coming"' (Bird Rose, 2008: 157). It alludes to a world of geographical, historical and ecological relationships which exists beyond possession and is instead about 'belonging' with them. Perhaps this relational sensibility is why colonial governments, obsessed with ownership, found it easy to dismiss Aboriginal presence and belonging and take possession of the land.

Ideas of land improvement in the colonies were not only expressed through the introduction of economically useful plants and animals, but also through that ideology of property, which requires wresting every bit of value out of every measure of land owned or occupied. Settlers were required to clear land for agriculture to prove productive use and show efficient stocking rates of sheep and cattle to secure pastoral leases on Crown Land. The gospel of efficiency justified land use practices that, over time, led to degraded soils and vegetation. Cycles of drought and floods added to land damage on a large scale. The conjuncture of greater environmental awareness, falling productivity and increased export competition, and recurring drought during the 1980s, motivated the Australian Commonwealth Government to reorient the discourse of land management from 'use' to land 'care'. Following the launch of the Victorian Landcare initiative in 1986, the Federal Government adopted the concept and founded Landcare at a national scale in 1989. Landcare was conceived as both an organisation and approach that would instil an ethic of 'stewardship' of the land among farmers and pastoralists and educate them to become 'adaptive resource managers' (Curtis and de Lacy, 1995). Farmer groups were to be mobilised for collective action centred on environmental repair and sustainable land use. A large part of the environmental repair involved eradication of invasive plant species.

By the end of the 1990s, the link between environmental repair, biodiversity protection, and agricultural productivity came to be predominantly centred on the threats posed by 'exotics' or 'alien invasives'. Reports on pastoral land management declared that 'the sustainability of the grazing industry is under serious threat from the invasion of exotic woody weeds' (March, 1995). In 1999, the Commonwealth Ministers of Agriculture, Forestry and Conservation, and Environment jointly announced the inaugural list of top 20 Weeds of National Significance (WoNS) that posed a threat to the Australian environment and which would receive priority funding for eradication programs. All the plants on the list were, without exception, 'exotics' and 'aliens' that had arrived during previous eras of experimentation with acclimatisation and land improvement, including the prickle bushes that had been introduced in Outback Queensland to aid the sheep industry.

A subtle shift in the Landcare discourse began to occur during the first decade of the twenty-first century. Holmes (2011) argues that this was largely due to the government's need to acknowledge the Australian High Court's recognition of Aboriginal Native Title to land: 'In this context, it became impossible for Landcare to continue to overlook Aboriginal people as both land managers and the traditional owners of the land.' (ibid.: 235). She points out that the discourse changed to both acknowledge indigenous sensibility and simultaneously appropriate it to reinforce settler-descendant narratives of stewardship and land ownership. Landcare became the symbol of restoring Australian landscapes to their earlier, pre-acclimatised state by removing alien invasive plants and replacing them with appropriate native species. Holmes quotes Kylie Mirmohamadi, who describes the enthusiastic adoption of native species in Australian gardens as 'effecting a recuperation of the landscape; purging the space of undesirable plants has restored it to its pre-lapsarian (in Australia, a pre-invasion) state of grace' (ibid.: 237).

In 2008, the Australian Government initiated a broad natural resource man-agement program called *Caring for **Our** Country* (bold in original). Planned to operate over a five-year period, *Caring for **Our** Country* integrated a number of land management programs such as National Landcare, Environmental Stewardship and Indigenous Land and Sea Rangers, which all involved, to varying extents, invasive species eradication and biodiversity protection (Commonwealth of Australia, 2013). The title of the broader program was adapted from the Northern Land Council's (NLC) 'Caring for Country' unit, launched in 1995 to help Aboriginal owners to manage their land and coastal areas. As Holmes (2011) observes, the Australian government's program is significant in not just appropriating the name of the NLC program, but reasserting a settler-descendant claim of ownership and national patriotism over land. 'Caring for Country' conveys the Aboriginal sensibility of 'country' as encompassing a living world of relationships and affinities. *Caring for Our Country*, on the other hand, conveys the sense of tending a possession. Despite these critical differences in meaning, the emergence of the word 'country' in contemporary land management discourse has revived debates regarding the role of native and introduced species in sustaining Australian agriculture, pastoralism, 'authentic' nature, and national identity.

In the following sections, we trace this evolution in thinking about 'country' through the history of pastoralism in Outback Queensland. We illustrate our discussion of contemporary concerns about sustaining the pastoral economy with interviews we conducted with pastoralists to elicit and compare their attitudes to prickle bush eradication and land management for cattle production.

Evolution of pastoralism in Outback Queensland

During the 1840s, the colonial governments of Victoria, New South Wales and South Australia were keen to find an overland route to the Gulf of Carpentaria so as to reduce shipping costs and garner greater profits from direct trade with the East India Company (Mitchell, 1848). Queensland was gradually opened through exploration and increasing settlement from the 1860s onwards. Much of the exploration and building of infrastructure was achieved by using cameleers and camels recruited from the northwest regions of the Indian subcontinent (Stevens, 1989). By the 1880s, most of the inland frontier was claimed as Crown Land by the colony and subdivided into pastoral leases.

Early pastoralism in northwest Queensland primarily centred on sheep rearing for wool. Remote sheep stations received supplies from and transported their wool to railheads and coastal ports using cameleers and their camels (Rangan and Kull, 2010). Pastoralists were largely concerned with rearing a healthy sheep stock that would produce wool of certain quality, shearing the sheep and shipping the wool off to market. Stocking rates on pastoral leases were partly determined by the colonial government which, ever committed to the ideal of 'improvement', specified the minimum number of animals to be grazed per unit area based on the size of their leasehold properties and availability of fodder and water (Seddon, 2003: 36). Sheep were left to graze across the properties and rounded up only during the shearing season or during periods of drought, when fodder and water were in short supply and had to be provided to keep the animals alive. This mode of pastoralism was significantly different from traditional Old World practices of raising sheep. For as long as pastoralism has been practised, it has been the shepherd's role to decide where and for how long to graze animals in particular areas before moving them on to other pastures. Shepherds were employed in early settler pastoralism in New South Wales and Victoria, but by the time the sheep industry developed in Queensland, they were no longer used. Moving stock from one pasture to another according to seasons, availability of fodder, water and shade was replaced by what might now, with hindsight, be called 'uncontrolled grazing'.

One of the biggest problems faced by pastoralists in northwest Queensland was the high mortality of lambs during the hot, dry season. The native grasses died back and native trees provided little feed or shade for animals to survive in the searing heat. Pastoralists thus sought to introduce palatable tropical grasses and hardy leguminous tree species that would provide the necessary shade and nutrition for the animals across their properties. During the 1890s, the Queensland Acclimatisation Society procured prickly acacia (*Acacia nilotica*) seeds from the British Indian colonial government's botanical gardens in Saharanpur and distributed it as a shade and forage tree for pastoral stations. The Queensland government, along with the Brisbane Botanical Gardens and individual botanical enthusiasts, embarked on a 'tropical acclimatisation [and improvement] experiment' by corresponding with plant and seed suppliers from tropical colonies to obtain, cultivate and disseminate useful plants to pastoralists in the colony (de Lestang, 1939; March, 2007). By the 1930s, prickly acacia was well established across much of northwest Queensland, and was celebrated for its ability to survive the hot season, and provide shade and protein-rich fodder for sheep. Other prickle bushes such as Parkinsonia (*Parkinsonia aculeata*, from Central America and the Caribbean), and mesquite (*Prosopis spp.*, also from Central America) were also introduced during the late nineteenth century. Parkinsonia was initially planted as an ornamental bush to provide shade and serve as a hedge around homesteads and watering points (Deveze, 2004). Mesquite was used as a fodder and shade plant in pastoral stations and also planted extensively around mining sites to minimise dust and soil erosion (Osmond, 2003).

The first half of the twentieth century proved fairly difficult for pastoralists as global wool prices experienced several dramatic collapses. Some began to convert their livestock from sheep to cattle for beef production. Until then, pastoralists in northwest Queensland had not raised cattle on any significant scale for export, but run them in areas where they knew sheep would not survive (Wadham, 1931). By the 1950s, the conversion of stock from sheep to cattle was almost complete and linked to growing domestic and international markets for beef. Cattle were subsumed into the same kind of pastoral system as sheep, with similar logic underlying stocking rates and grazing practices. They grazed across the property and were mustered a few times in the year to be taken to market or when there was serious fodder shortage during droughts. Cattle became the iconic symbol of pastoral life in Outback Queensland and the grand narratives of rugged Australian settler identity. Similar to the western USA where the cattle industry symbolised the tough frontier spirit of cowboys, the Queensland Outback represented the remote inland and northern frontier of Australia where pastoralists, jackaroos, and drovers worked in harsh, unforgiving environments to muster the animals and take them on long stock routes from inland stations to coastal stockyards.

Pastoralism in Queensland underwent another significant shift between the 1950s and 1970s, this time in the organisation of production. In 1959, a report prepared by the national Department of Agricultural Economics recommended the reorganisation and expansion of the beef industry in northwest Queensland (Kelly, 1959). It called for a fundamental restructuring of the beef industry to promote greater efficiencies and economic advantage by separating and specialising in different stages of production and value chain. Instead of cattle being born, raised and fattened on a single property before being sent to market, the report recommended that pastoralists could specialise in different stages of the animal's life cycle depending on the resource attributes and locational advantages of their properties; some could specialise in breeding, others could raise cattle to maturity, and yet others could specialise in fattening or 'finishing' the cattle on richer pastures or feedlots before they were sent to slaughter. Improvements in road infrastructure combined with innovations such as large 'road trains' for transporting cattle *en masse* from one production site to another along the beef value chain also enabled quicker returns and the promise of larger profits for pastoralists. Although the proposal seemed fairly radical in the 1960s, this 'flexible industrial' system has since become the norm for large-scale cattle production in Queensland and other regions and countries involved in global beef markets and commodity chains. The pastoral economy of northwest Queensland was thus transformed from a grazing-centred industry to industrial pastoralism requiring greater capital investment in technology and infrastructure along the commodity chain.

Most pastoral properties in northwest Queensland realigned and integrated with the new system of beef production as domestic and global demand for beef grew through the 1970s. The landscape, too, reflected the shift from sheep to cattle and presented new problems for pastoralists at the early stages of the value chain. Following the heavy rains and floods in 1974, pastoralists confronted a massive expansion of prickle bushes that had previously dotted the grazing landscape and lined bore drains. None of them recollected previous floods as having caused such an explosion of prickly trees. One pastoralist recalled:

> We'd started off with two trees at 'Isobel'. We thought they were wonderful. My husband and I were planting them along the bore drains and everything . . . In 1974 the prickly acacia grew out of the floods in all proportions . . . it went on from September that rain, from September to March. It rained in April and those seeds got absolutely soaked . . . and we didn't realise that we had such a terrible problem . . . and after '74 I just hated that tree because it infested our place.
>
> *(Pastoralist E, interviews, 2007)*

One of the main reasons for the proliferation of prickle bushes in pastoral holdings was the replacement of sheep with cattle. Although prickle bushes still served the purpose of providing shade and fodder for the animals, they behaved differently in a landscape grazed by cattle instead of sheep. As far back as 1926, the *Queensland Agricultural Journal* observed that *A. nilotica* was likely to spread more vigorously with cattle grazing, noting:

> There, is, however, a drawback to this tree in cattle country, in that cattle consume the pods, the seeds are not masticated and pass whole through the digestive tract, thus causing numbers of young trees to appear where they are not wanted.
>
> *(quoted in Spies and March, 2004)*

Several other studies have supported this observation, noting that cattle and sheep graze and process hard seeds differently. Sheep chew leguminous pods with hard (acacia) seeds to a point that makes them unviable for germination when passed through droppings. Cattle do not break down seed pods in the same way, which results in more undigested seeds passing through the animal and remaining in the dung until appropriate conditions for germination arise (Simao-Neto *et al.*, 1987; Tiver *et al.*, 2001).

The second reason for the growing condemnation of prickle bushes was the shift to more machine-reliant methods of mustering cattle. Mustering has long been the most iconic representation of Outback Australian life, an activity that evokes enormous pride and sense of identity for pastoralists in the region (March, 1995). In the past, mustering was a labour-intensive logistical operation involving a large number of stockmen on horseback, who rounded up cattle for drovers who would to take them to stockyards for sale. But the new industrial pastoralism has largely replaced horses with quad-bikes, motorbikes, 4WD vehicles, helicopters and small aircraft for mustering cattle. Many pastoral properties have invested in such equipment to manage large herds and muster cattle with fewer people. These new machine-based mustering methods are more affected by the spread of prickle bushes than mustering on horseback. Vehicle tyres experience substantial damage from the long and sharp acacia thorns, and aerial mustering becomes more difficult when cattle run into and hide in dense thickets of prickle bush. It is in this context that prickle bushes have become the villains of pastoralism and targeted as alien invasive species that need to be eradicated from the landscape.

Tackling prickle bushes

> An old farmer went to a Landcare conference, an old fella, about 75, and he went along and he stood up at the end of that conference and he said: 'I just hope this Landcare works because I'm sick to death of poisoning the things that want to live here and trying to raise the things that want to die.'
>
> *(Pastoralist A, interviews, 2007)*

The rise of industrial pastoralism has added to the many pressures that affect the viability of family-run pastoral businesses in northwest Queensland. Globally linked agribusinesses have taken over management of vast tracts of land under pastoral leases along the beef production chain and have invested capital in large-scale mustering and transport logistics. The technological composition of capital in the landscape and production is increasingly defined by such companies and family-run pastoral businesses need to achieve comparable levels of efficiency to remain profitable. In this context, prickle bushes have come to symbolise not just a weedy plant but an entire complex of problems for pastoral families in the region. Not all have the capital to use the technologies that large agribusinesses may use to clear prickle bushes from their properties. Not all pastoral family businesses have the capacity to keep abreast of various environmental regulations governing vegetation clearance or the resourcefulness to garner funds from Landcare, WoNS, and other government programs for eradicating prickly acacia. For many, the prickle bushes are the new enemy invading their land and which need to be defeated in battle.

The main government agency interacting with pastoralists on the eradication of alien invasive plants is the Queensland Department of Agriculture, Forestry and Fisheries (DAFF).[1] The agency provides support and advice to pastoralists on matters ranging from choice of pasture grasses, quarantine, drought mitigation and water, to weeds and pest management. The National Prickly Bush Management Group operates in conjunction with this department and other state agencies for Water, Natural Resources, Parks and Wildlife, to fund and assist pastoralist groups who are active in Landcare programs targeted at eradication of WoNS, so as to improve sustainability of the grazing industry and preserve native biodiversity (March, 1995). The normal practice of eradication usually involves a combination of physical uprooting and application of chemical herbicides to destroy the plants (DPI Official, interviews, 2007).

We explored the differing perspectives regarding prickle bushes in northwest Queensland by interviewing a number of family-run pastoral business owners. We asked them about their experiences of pastoralism, how they managed their land, what approaches they used for eradicating prickle bushes, and the challenges they faced in the industry.[2] Based on these interviews, we elicited four perspectives that broadly represented the concerns and approaches of family-run pastoral business owners regarding land and prickle bush management. We have classified the owners as: *Pragmatists*, *Unsuccessful Battlers*, *Resisters* and *Strategists*, and describe their views below.

Pragmatists[3]

A number of pastoralists pragmatically adopt the official Landcare perspectives regarding land management and use government agency grants and resources to eradicate prickle bush from their properties. Members of one family, well known in the region as successful battlers against prickle bush (Spies and March, 2004), declared that 'prickly acacia is the greatest threat the Artesian Basin has ever seen'. They talked about the presence of these trees as emblematic of land that was struggling to be productive: 'There is nothing more demoralising than seeing your country struggling you know, after it has been droughted or strangled with those trees' (Pastoralists C, B, S, interviews, 2007). Once they decided that this could not go on, they mobilised all their resources to free the land from prickle bushes. Over a period of ten years, they used chemicals, bulldozers and aerial spraying to eradicate the trees. This was done by members of the family with help from paid labour and labour 'barter days' with neighbours. Large thickets were sprayed by helicopter, widely dispersed individuals removed by quad bikes, and others were uprooted using bulldozers. The son, who has taken on the project as his personal mission, believed that, 'The fight is a long-term thing.' Pointing to an area of grassland that had previously been a forest of prickly acacia, he declared, 'I think that if we keep going for another ten, twenty years we may finally see the seed bank disappear' (Pastoralist S, interviews, 2007).

Unsuccessful Battlers

Not all pastoralists were as successful as the Pragmatists in clearing prickle bush from their land. Older pastoralists, who saw themselves as being at the forefront of land improvement and forging the pastoral industry in the region, were disheartened by the difficulties of clearing the trees from their land. They pointed out that their life experiences, local knowledge of prickly tree behaviour, and understanding of the physical character of their pastoral properties were often ignored by the 'best practice' approach advocated by government agencies for eradicating the trees. They knew from experience that chemical application and uprooting did not affect large seed banks of prickly acacia in the soil and that new seedlings would spring up in profusion following the wet season. They were also frustrated by the contradictory policies and regulations that worked against prickly acacia clearance (Pastoralists G, J, L, Ja, Sa and N, interviews, 2007). One of the chief sources of irritation was the Queensland Vegetation Management Act, 1999 (VMA). This Act was instituted following widespread concern over the rapid rate of vegetation and tree clearance and conversion of freehold land for potentially unsustainable uses. It rules against removal of native tree species and requires landholders to obtain permits to clear trees – even those deemed non-native or invasive – from riparian areas (Queensland Government, 1999). An older pastoralist claimed he was flabber-gasted at the stupidity of the Act which prevents landholders from using tractors and bulldozers within 20 metres of waterways in their properties, and requires permits to clear declared weeds even outside the waterway boundaries. 'We are trying to eradicate it, that's the goal,' he explained, 'but how's this? It took four months to get a permit to kill a declared weed, and by this time the permit was a complete waste of time as [the weed] spread rapidly and destroyed half the property while we waited' (Pastoralist G, interviews, 2007).

Many of these Unsuccessful Battlers are of an older generation that remembers when government agencies vigorously encouraged the planting of prickly acacia in the past as much as they vigorously advocate its eradication in the present (Pastoralists G, J, C, Sa and N, interviews, 2007). They pointed out that the idea of good stewardship of the land was constantly changing and this placed them in a situation of permanent struggle to manage their land and business and make a viable living. Some, like the 'old fella' mentioned in the quote at the beginning of the section, felt they no longer had the energy or will to undertake Landcare or other government programs that claimed to protect native biodiversity and improve sustainability of the pastoral industry. Others wondered whether the so-called best practices of Landcare and government policies were merely using the trendy language of sustainability or genuinely 'doing right by country' (Pastoralist A, interviews, 2007).

Resisters

Given the broad ethos of antagonism toward prickly acacia, very few pastoralists in the region wanted to be seen as advocates for the tree. One pastoralist was an exception in resisting the pressure to condemn prickly trees as evil. 'I wouldn't be here if it wasn't for my prickly trees,' he declared. He claimed that the prickly acacias were the best source of protein for sheep and cattle available in this part of the Outback and were crucial for keeping them healthy during good and bad times. He explained:

> Round here at the end of every year when they got beans on them . . . it's like a measure of protein drops on the ground every day . . . The sheep come along and they get their feed on the protein and it keeps them going, you know, in the worst of the drought.
>
> *(Pastoralist D, interviews, 2007)*

This pastoralist took great pleasure in poking fun at government programs and Landcare initiatives for their fear-laden concerns about the threats posed by prickly acacia and insisted on celebrating their role in making pastoralism successful in the region.

Strategists

Another group of pastoralists who refused to condemn prickle bushes stood out from the rest by emphasising the broader aspects of pasture management. They claimed that it had become a habit to constantly look for scapegoats to blame for all the problems associated with pastoral production. The prickly acacia was the latest target of such scapegoating, which led many people to think that all their productivity and industry-related problems would be solved by eradicating these trees. They argued that, unlike the conventional perspective of seeing land as a resource that exists to support cattle production, their main focus was on producing pastures with healthy soil and grass. Rather than being cattle graziers, they saw themselves as pastoralists who harvested 'sunlight, grass, and soil' embodied in the form of cattle.

The strategic shift in thinking for this group of pastoralists stems from their espousal of an alternative approach developed by a private business called the Grazing for Profit (GfP) school. This school draws on the ideas and philosophy of Alan Savory, a pastoralist from Zimbabwe, who developed the 'cell grazing' technique based on his observations of wild ungulate behaviour on the savannahs of southern Africa. One member described cell-grazing as 'an intensive system of rotational grazing, utilising the tools of stock density, feed utilisation, and rest to enhance natural ecosystems' (Pastoralist A, interviews, 2007). Simply put, it means that instead of leaving cattle to graze extensively on large paddocks, they graze intensively for short periods of time across a number of smaller paddocks. The cell-grazing system is not unlike what traditional herders and shepherds have done in Africa and elsewhere. However, this new conceptualisation of traditional herding as cell-grazing has gained purchase with a number of pastoralists in Australia and the USA who had previously practised 'uncontrolled', extensive grazing.[4]

The need for strategic redirection for these pastoralists arose because they found their businesses were becoming increasingly unviable. They were 'just surviving' or had reached the point of being 'sick of constantly battling against the land' (Pastoralists A, K, Ji and T, interviews, 2007). When they looked for alternative ways of improving their business and encountered the GfP approach, they felt that it provided a new framework which interlinked environmental, social, economic and personal challenges and made them realise that 'we need to stop blaming the governments and the weather and look at managing what we have' (Pastoralist A, interviews, 2007). The GfP approach resonated with them because they could use Savory's metaphor of the landscape as a continuum ranging from brittle to non-brittle, to match observations of soil and grass quality on their properties and use this information to fine-tune their pasture and grazing strategies. Some members explained the new approach to 'grass farming' by drawing parallels with raising lawns, using cattle as natural lawnmowers and fertilisers. Although cell-grazing is more labour-intensive than extensive grazing due to the need for animals to be moved every three or four days depending on grass growth in each paddock, the technique worked well alongside other strategies such as Low Stress Stock Management (LSS), which are aimed at producing healthier cattle for higher quality beef. One respondent said she regularly 'walked the weaners' so that they felt comfortable with people and dogs while being moved from one paddock to another (Pastoralist K, interviews, 2007). They noted that cell-grazing and LSS enabled them to think differently about producing healthy pastures and healthy animals from existing landscapes (Pastoralists A, K, Ji, T, and Jo, interviews, 2007). One pastoralist who was an avid spokesman for the approach described the benefits on a national television program, saying:

> I can see just the change that comes upon us when we make some mental changes in our attitude . . . what we are doing changes . . . And it's pretty rewarding when you can work with the land and the animals and the people in harmony.
>
> *(Lindsay, 2001)*

Prickle bushes were not the object of concern for these pastoralists. One declared that it was futile to focus on eradicating prickly acacias in conventional ways because it would effectively mean wearing 'a chemical company's hat and advertise for someone who takes all your profits'; the companies were only interested in 'selling a farmer a chemical so that he can go and poison prickly bushes for the next hundred years and still not beat 'em' (Pastoralist J, interviews, 2007). A few had taken a different approach to managing prickle bush in grazing areas by bringing cattle and camels together for cell-grazing. Camels, also a legacy of outback frontier settlement and bigoted White Australia policy (Baker, 1964; de Lepervanche, 1984; Stevens, 1989),[5] have been classified as feral by government agencies and routinely targeted for mass culling in the Outback areas where they were released during the 1920s and 1930s (Rangan and Kull, 2010, see also Chapter 2 in this volume). These pastoralists have succeeded in using the camels on their properties to graze with cattle. Camels prefer browsing on prickle bushes and break down the hard seeds far more thoroughly in their digestive system than cattle. The pastoralists claim that this co-grazing technique combined with cell-grazing has proved more effective in maintaining good pastures with less-dense clumps of prickle bushes and has produced healthy cattle (Pastoralists JL, JW, TL, interviews, 2007; Phillips *et al.*, 2001; Dörges and Heucke, 2003).

The approach adopted by these 'strategic' pastoralists is to begin by accepting the landscape for what it is and working with it to make it healthy. They believe their techniques require a greater understanding of the dynamic environmental and ecological conditions of interactions between animals, grass, bush, and soil on their land and a perspective that focuses on 'doing right by country' rather than 'battling' the land or prickle bushes or camels or any other 'invasive' species. They feel this has given them greater confidence for remaining profitable in the pastoral industry.

Conclusion

Pastoralism and prickly acacia have co-evolved in the landscapes of northwest Queensland. Both were introduced, were not native to the region. The shade and fodder offered by the prickly acacia in the first half of the twentieth century increased the viability of the landscape for sheep and helped pastoralists remain profitable. As pastoralism evolved from sheep to cattle, so did the landscape, along with prickle bushes. The shift to industrial pastoralism has increased the pressure on family-run pastoral businesses to operate with the same technological capacity as large-scale agribusiness companies, and pastoralists have responded in different ways to these pressures. Few feel that eradicating prickle bushes can solve all the other problems they face in remaining profitable in the industry. Since we did our fieldwork, repeated flooding during the wet season over the past five years in northwest Queensland has, without doubt, created extremely favourable conditions for prickle bush to expand. 'Battling' this expansion on a similar scale is something that few family-run pastoral businesses can afford to take on. The shifting global economic geography of the beef industry also imbues family-run pastoral businesses with a greater sense of uncertainty regarding their futures. They realise that, just as happened with the wool industry during the first half of the twentieth century, fluctuations or collapses in global commodity prices for beef may well require them to reorient their businesses to some other land-based activity. Under such conditions of uncertainty, the best strategy, as some pastoralists have shown, is to focus on understanding the dynamic landscape and its varied biological components and processes so as to keep the 'asset base' in healthy condition.

Our chapter also points to a gradual change occurring in the native and alien species debate in Australia. The 'strategic' pastoralists demonstrate a marked paradigm shift in refusing to buy into the labelling of 'good' and 'bad' plant species, native or non-native. Among those involved in invasive species debates within both academic and policy circles, there is greater awareness of the substantial critique of the concepts and metaphors used in invasion biology from both within and outside the discipline (Comaroff and Comaroff, 2001; Subramaniam, 2001; Larson, 2005; Davis, 2009; Davis et al., 2011). Concepts of 'novel' and 'hybrid' ecosystems (Hobbs et al., 2006; Smith et al., 2006; Davis, 2009) have given rise to discussions of how dynamic urban and range ecologies might be managed for greater biodiversity. However, the ways in which these new theories are adapted to policy will require a renegotiation of the nationalist narratives of landscape, particularly in Australia and other settler-immigrant nations. There will always be pressure to articulate policies in terms of 'biosecurity' and the threat of 'aliens' landing from foreign shores. Instead of acknowledging that ecologies and landscapes change in relation to economic shifts and socio-cultural values, these policies are likely to blame particular biological species for causing problems and target them for eradication.

There is also a subtle, yet profound, change in sensibility taking place among some of the pastoralists towards land. The government's reconfiguration of Landcare programs under the slogan 'Caring for **Our** Country' has, to some extent, drawn attention to the tensions arising from bringing particular notions of 'care' and 'country' together. The 'battler' narrative sits awkwardly within this con-ceptualisation and many pastoralists questioned it in different ways as they reflected on ways of looking after their properties and making a living from pastoralism in the future. They were less concerned about land 'care' and its multiple philosophical meanings and more about 'doing right *by* country'. The expression 'doing right' signals a sense of respect and duty towards someone else, in this case towards 'country', without reference to possession. Their use of 'country' in the Aboriginal English sense of the term may well lead to new pastoral and Outback landscapes that reflect the sensibility of 'doing right by country'.

Acknowledgements

We thank the Australian Research Council for funding this research, which was part of a larger project entitled, 'Acacia exchanges around the Indian Ocean' (DP0666131), led by Christian Kull and Priya Rangan, and provided scholarship and fieldwork support for Anna Wilson. We thank Pat Lowe, Judith Carney, Heather Goodall, Jodi Frawley, Virginia Brunton, Lesley Head, and many more participants of the Rethinking Invasion Ecologies in the Anthropocene conference/ workshop at the University of Sydney for their insights and comments on the ideas presented in this chapter.

Notes

1   This agency has been renamed several times, depending on the government in power. Over the past ten years, it has been called Department of Primary Industries (DPI), then renamed Department of Employment, Economic Development, and Innovation (DEEDI), and is currently called DAFF (Queensland Government, 2013). Essentially, it is the department that is responsible for overseeing the state's agricultural and pastoral sectors.

2   Fieldwork was initiated by Christian Kull and Priya Rangan as part of their ARC-DP 0666131, 'Acacia exchanges around the Indian Ocean'. They set up meetings with government officials and conducted a preliminary survey of the areas targeted for prickle bush control in the Southern Gulf Catchment area of northwest Queensland. Anna Wilson and Alyse Weyman were recruited as Master's research students to develop their thesis projects within the rubric of the larger project, and to carry out field interviews with pastoralists in this region. They interviewed family-run cattle stations within 100 km of Flinders Highway. The pastoralists were selected through a snowball process, beginning with a contact list provided by key representatives of the state Department of Primary Industries (DPI) and the National Prickly Bush Management Group. The agency officials suggested a range of contacts with different ideas and approaches towards land, prickle bushes and pastoral management. Semi-structured conversations were carried out with 35 individuals associated with family-run properties, and with 34 individuals from 19 organisations and government departments. Interviews with pastoralists were conducted on their stations both in their homes and while touring their properties.

3   Many thanks to Virginia Brunton for pointing out to Priya Rangan that this group should be categorised as *Pragmatists* because they utilised the funding opportunities for prickle bush clearance provided through Landcare and WoNS programs.

4   Savory's cell-grazing technique has been positively evaluated by some scientists (Wolf and Allen, 1995; Gompert, 2006; Roncoli *et al.*, 2007), while others claim the benefits are uncertain and scientifically inconclusive (Brown, 1994; Dagget and Dusard, 2000; Li *et al.*, 2002). Savory's 'holistic' approach to pastoralism is similar to that of traditional pastoral communities that recognise the role of animals in creating and using their ecosystems by fertilising and nourishing the grasses and soils they need to thrive. The holistic agricultural approach was also emphasised by Albert Howard during the 1940s, who called on farmers to take a step back and watch and learn from nature, 'the supreme farmer'. He argued that industrial agriculture generated enormous waste in production and undermined the longer-term health of soils and ecosystems. The healthiest way to farm was to mimic nature and ensure that nothing was wasted; animals and plants needed to be managed in rotation between fallow and productive areas in ways that preserved soils and enriched the farming ecosystem (Howard, 1943). Howard's ideas are better known within the contemporary organic farming movement than among the pastoralists we interviewed.

5   The history of camels and prickly trees in the making of Outback Australia is described in greater detail in Rangan and Kull (2010). Following the passage of the Immigration Restriction Act of 1901, the Afghan cameleers were not given citizenship because of their ethnicity. Many of them released their camels in the Outback before leaving for the Indian subcontinent. Just as the shift in the pastoral economy from sheep to cattle led to a landscape with more prickle bushes, the White Australia policy led to an Outback with feral camels. In this sense, feral camels can be seen as living reminders of the forging of

White Australian national identity and as much a part of the Australian Outback as kangaroos and emus.

References

Anderson, K. 2003. White natures: Sydney's Royal Agricultural Show in post-humanist perspective. *Transactions of the Institute of British Geographers*, 28, 422–441.

Anon. 1866. The Acclimatisation Society. *Brisbane Courier*, 6 October, p. 5.

Baker, H. 1964. *Camels and the Outback*, Melbourne, Pitman & Sons.

Bean, C. 1910. *On the Wool Track*, London, A. Rivers.

Bird Rose, D. 2008. On history, trees, and ethical proximity. *Postcolonial Studies*, 11, 2, 157–167.

Brown, D. E. 1994. Out of Africa. *Wilderness*, 58, 24–33.

Clarke, M. 1909. Preface. In: Lindsay Gordon, A. *Sea Spray and Smoke Drift*, Melbourne, Lothian, p. iii.

Comaroff, J. and Comaroff, J. 2001. Naturing the nation: aliens, apocalypse and the postcolonial state. *Journal of Southern African Studies*, 27, 627–651.

Commonwealth of Australia, *Caring for Our Country*. Available at: www.nrm.gov.au/about/caring/ (accessed 15 June 2013).

Curtis, A. and De Lacy, T. 1995. Examining assumptions behind Landcare. *Rural Society*, 5(2–3), 44–55.

Dagget, D. and Dusard, J. 2000. *Beyond the Rangeland Conflict: Toward a West that Works*, Reno, Nevada, University of Nevada Press.

Davis, M. 2009. *Invasion Biology*, New York, Oxford University Press.

Davis, M. A., Chew, M. K., Hobbs, R. J., Lugo, A. E., Ewel, J. J., Vermeij, G. J., *et al.* 2011. Don't judge species on their origins. *Nature*, 474, 153–154.

Davison, G. and Brodie, M. (eds) 2005. *Struggle Country: The Rural Ideal in Twentieth Century Australia*, Melbourne, Monash University Press.

de Lepervanche, M. M. 1984. *Indians in a White Australia: An Account of Race, Class and Indian Immigration to Eastern Australia*, Sydney, George Allen & Unwin.

de Lestang, A. 1939, Letter to D.M. Gordon, Glenmorgan, Adel's Grove Co.

Deveze, M. (ed.) 2004. *National Case Studies Manual – Parkinsonia: Approaches to the Management of Parkinsonia (Parkinsonia aculeata) in Australia*, Brisbane, Government of Queensland.

Dörges, B. and Heucke, J. 2003. *Demonstration of Ecologically Sustainable Management of Camels on Aboriginal and Pastoral Land*, Final report on project number 200046 for the Natural Heritage Trust, Alice Springs, CACIA.

Gompert, T. 2006. Managed grazing in sustainable farming systems. In: Francis, C. A., Poincelot, R. P. and Bird, G. W. (eds) *Developing and Extending Sustainable Agriculture: A New Social Contract*, Philadelphia, PA, Haworth Press, pp. 65–89.

Hobbs, R. J., Lugo, A. E., Norton, D., Ojima, D., Richardson, D. M., Sanderson, E. W. *et al.* 2006. Novel ecosystems: theoretical and management aspects of the new ecological world order. *Global Ecology and Biogeography*, 15, 1–7.

Holmes, K. 2011. Redeeming landscapes: Ireland and Australia. In: Holmes, K. and Ward, S. (eds) *Exhuming Passions: The Pressure of the Past in Ireland and Australia*, Dublin, Irish Academic Press, pp. 223–242.

Howard, A. 1943. *An Agricultural Testament*, Oxford, Oxford University Press.

ISSG. 1995. *Aliens Newsletter*, 1, 13–14.

Kelly, J. H. 1959. *The Beef-Cattle Industry in the Leichhardt-Gilbert Region of Queensland*, Canberra, Bureau of Agricultural Economics.

Larson, B. 2005. The war of the roses: demilitarizing invasion biology. *Frontiers of Ecology and the Environment*, 3, 495–500.

Li Y., Johnson, D. L. and Marzouk, A. 2002. Pauperizing the pastoral periphery: the marginalisation of herding communities in the world's dry lands. *Journal of Geographical Sciences*, 12, 1–14.

Lindsay, J. 2001. Speak softly to wield a big herd. *Landline*, ABC Radio.

March, N. 1995. *Exotic Woody Weeds and their Control in North West Queensland*, Cloncurry, Queensland Department of Natural Resources, Mines and Energy.

March, N. 2007. *Prickly Acacia, Parkinsonia and Mesquite National Distribution*, Cloncurry, Queensland Department of Natural Resources, Mines and Energy.

McGuire, P. 1939. *Australia, Her Heritage, Her Future*, New York, Frederick Stokes & Co.

Mitchell, T. L. 1848. *Journal of an Expedition into the Interior of Tropical Australia: In Search of a Route from Sydney to the Gulf of Carpentaria*, London, Longman, Brown, Green and Longmans.

Muller, S., Power, E. R., Suchet-Pearson, S., Wright, S. and Lloyd, K. 2009. 'Quarantine matters!': Quotidian relationships around quarantine in Australia's northern borderlands. *Environment and Planning A*, 41, 780–795.

Osmond, R. (ed.) 2003. *Best Practice Manual – Mesquite: Control and Management Options for Mesquite (Prosopis spp.) in Australia*, Brisbane, Government of Queensland.

Phillips, A., Heucke, J., Dörges, B. and O'Reilly, G. 2001. *Co-Grazing Cattle and Camels: A Report for the Rural Industries Research and Development Corporation*, Canberra, RIRDC.

Powell, J. 1988. *A Historical Geography of Modern Australia: The Restive Fringe*, Cambridge, Cambridge University Press.

Queensland Government. 1999. *Vegetation Management Act (VMA) 1999: An Act about the Management of Vegetation on Freehold Land*, Brisbane: Office of the Queensland Parliamentary Counsel.

Queensland Government. 2013. *Weeds*. Available at: http://www.daff.qld.gov.au/26_8331.htm (accessed 15 June 2013).

Rangan, H. and Kull, C. 2010. The Indian Ocean and the making of outback Australia: an ecocultural odyssey. In: Moorthy, S. and Jamal, A. (eds) *Indian Ocean Studies: Cultural, Social, and Political Perspectives*, New York, Routledge, pp. 45–72.

Robin, L. 2007. *How a Continent Created a Nation*, Sydney, UNSW Press.

Roncoli, C., Jost, C., Perez, C., Moore, K., Ballo, A., Cissé, S., and Ouattara, K. 2007. Carbon sequestration from common property resources: lessons from community-based sustainable pasture management in north-central Mali. *Agricultural Systems*, 94, 97–109.

Seddon, G. 2003. Farewell to Arcady: or getting off the sheep's back. *Thesis Eleven*, 74, 35–53.

Simao-Neto, M., Jones, R. M. and Ratcliff, D. 1987. Recovery of pasture seed ingested by ruminant 1: seed of six tropical pasture species fed to cattle, sheep and goats. *Australian Journal of Experimental Agriculture*, 27, 239–246.

Smith, R., Maxwell, B., Menalled, F. and Rew, L. 2006. Lessons from agriculture may improve the management of invasive plants in wildland systems. *Frontiers in Ecology and the Environment*, 4, 428–434.

Spies, P. and March, N. 2004. *Prickly Acacia National Case Studies Manual*, Cloncurry: Queensland Department of Natural Resources, Mines and Energy.

Stevens, C. 1989. *Tin Mosques and Ghan Towns: A History of Afghan Camel Drivers in Australia*, Melbourne, Oxford University Press.

Subramaniam, B. 2001. The aliens have landed! Reflections on the rhetoric of biological invasions. *Meridians*, 2, 26–40.

Tiver, F., Nicholas, M. and Kriticos, D. 2001, Low density of prickly acacia under sheep grazing in Queensland. *Journal of Range Management*, 54, 382–389.

Turner, I. 1968. *The Australian Dream: A Collection of Anticipations about Australia from Captain Cook to the Present Day*, Melbourne, Sun Books.

Tyrrell, I. 1999. *True Gardens of the Gods: Californian-Australian Environmental Reform, 1860–1930*, Berkeley, CA, University of California Press.

Wadham, S. 1931. The pastoral industries. *Annals of the American Academy of Political and Social Science*, 158, 40–48.

Wolf, S. and Allen, T. 1995. Recasting alternative agriculture as a management model: the value of adept scaling. *Ecological Economics*, 12, 5–12.

PART IV

Ecological politics of imagining otherwise

9

PRICKLY PEARS AND MARTIAN WEEDS

Ecological invasion narratives
in history and fiction

Christina Alt

In 1898, R. A. Gregory, the editor of *Nature*, published a review of H. G. Wells's
novel, *The War of the Worlds*, under the title 'Science in Fiction'. Gregory praised
Wells for his 'ingenuity in manipulating scientific material', noting Wells's
engagement with Percival Lowell's astronomical observations of Mars, his speculation regarding the evolutionary development and technological prowess of the
Martian invaders, and the 'distinctly clever' plot twist centring around the Martians'
susceptibility to earthly germs (Gregory, 1898: 339). Both the specific points of
scientific interest identified by Gregory and the fact that Gregory regarded Wells's
novel as warranting a review in the preeminent science journal of the period suggest
the extent of Wells's engagement with contemporary science.

Subsequent scholarly analysis of Wells's engagement with science in his fiction
has tended to cluster around the same topics that Gregory identified: astronomy,
evolutionary development, technology, and bacteriology. Roslynn D. Haynes's
H. G. Wells, Discoverer of the Future: The Influence of Science on His Thought (1980)
is a key text in this regard and demonstrates the value of reading Wells's work in
relation to the science of his time. More recently, critics have begun to consider
Wells's treatment of ecological themes, but here, in contrast to early reviewers like
Gregory and critics such as Haynes, most critics have stressed Wells's anticipation
of current ecological concepts and concerns rather than his engagement with the
science of his time. David H. Evans, for example, characterises Wells's account of
the Martian red weed as an 'uncanny premonition of certain recent examples of
imperialistic flora, like the kudzu vine' (Evans, 2001: 12).[1]

Wells's reputation for prescience is well deserved, and Evans's interpretation of
the Martian red weed warrants consideration, but it is also useful to examine the
extent to which Wells drew on contemporary scientific knowledge in his conception of this Martian vegetation. In an article on the history of invasion ecology
and changing perceptions of the tamarisk plant, Matthew K. Chew (2009) makes

reference to *The War of the Worlds*. He argues that Wells's Martian red weed was inspired by a particular historical model, the Canadian waterweed, thus illustrating the viability of a historical approach to Wells's representation of invasive species. This chapter will further develop this historical approach, tracing the understanding of introduced species that was in circulation at the time when Wells wrote *The War of the Worlds* and analysing the imaginative ends to which Wells put this scientific knowledge.

The War of the Worlds tells the story of the invasion of Earth by highly evolved, technologically advanced Martians trying to escape their cooling planet. The Martians come to Earth equipped with an arsenal of technological weaponry – heat rays, giant tripods, and digger machines – but they also bring with them Martian plant species which swiftly become pests: truly *alien* invasives. The Martian plants grow 'with astonishing vigour and luxuriance', spreading across the countryside in the space of days, leaving in some places not 'a solitary terrestrial growth to dispute their footing' (Wells, [1898] 2009: 150, 167).

Historical models for such plant incursions existed in the nineteenth century and many were well documented by the time Wells wrote *The War of the Worlds*. In his account of the voyage of the *Beagle*, published in 1839, Darwin describes three European plant species introduced to South America that had become 'extra-ordinarily common' there (Darwin, 1839: 119).[2] He records the growth of fennel 'in great profusion' in ditches around Buenos Aires, Montevideo and other towns and the spread of the giant thistle through the Pampas (ibid.: 119). Even more striking is his description of the thistle-like cardoon (*Cynara cardunculus*):

> In [Banda Oriental] alone, very many (probably several hundred) square miles are covered by one mass of these prickly plants, and are impenetrable by man or beast. Over the undulating plains, where these great beds occur, nothing else can now live. Before their introduction, however, the surface must have supported, as in other parts, a rank herbage. I doubt whether any case is on record of an invasion on so grand a scale of one plant over the aborigines.
>
> *(ibid.: 119)*

In *On the Origin of Species*, Darwin returns to the subject of introduced plants and to the examples of 'the cardoon and a tall thistle, now most numerous over the wide plains of La Plata, clothing square leagues of surface almost to the exclusion of all other plants' (Darwin, 1860: 65).[3] Darwin offers these examples of 'the extraordinarily rapid increase and wide diffusion of naturalised productions in their new home' to illustrate his argument that all species tend towards a geometrical rate of increase (ibid.: 65). In the process, he also furnishes a succinct explanation of the reasons behind the success of certain introduced species: placed in an environment in which all of its requirements for life are met and few limiting factors are present, a species will increase in its numbers at an exponential rate.

Darwin's account of the cardoon in South America almost certainly underpins – or underwrites – Wells's first description of the Martian red weed, and Wells's tale of Martian plant invasion can also be read as an elaboration of Darwin's characterisation of the spread of the cardoon through South America as an 'invasion' on a grand scale (Darwin, 1839: 119). However, in conceiving of the Martian red weed, Wells likely drew on more contemporary sources as well. *Nature* was a key source of up-to-date scientific information for Wells in this period.[4] Wells began publishing reviews, notes, and articles on science education in *Nature* in 1894, and the frequent presence of his work in the journal in the 1890s suggests that he was well acquainted with its contents in this period. In *The War of the Worlds*, he explicitly cites an article from the 2 August 1894 issue of *Nature* that described an unexplained light briefly observed on the surface of Mars. This astronomical observation was one of the inspirations for Wells's fictional tale of extra-terrestrial life and interplanetary invasion. In a note in the 6 May 1897 issue of *Nature*, a contributor draws attention to the serialised publication of *The War of the Worlds* in *Pearson's Magazine* and remarks, 'It is evident from many paragraphs that Mr. Wells reads his NATURE, and closely follows the planetary observations described in our astronomical column from time to time' (Anon, 1897: 16). Given these direct references to Wells's use of content drawn from *Nature*, it seems plausible to assume that discussions of introduced species in the journal may have shaped Well's conception of the Martian red weed.

From *Nature*, one can also trace a widening web of sources directly or indirectly available to Wells, for *Nature* regularly reproduced extracts from publications not easily accessible in Britain and recorded the contents of new issues of other influential journals, such as the *Kew Bulletin*. One relevant extract, reprinted in the 15 June 1893 issue of *Nature*, was an article on 'The New Flora and the Old in Australia' excerpted from a longer work first published in *The Journal and Proceedings of the Royal Society of New South Wales*. Written by A. G. Hamilton, the article focuses on the impact of introduced plants upon native Australian species and thus serves as an indicator of the late nineteenth-century understanding of introduced species that informed Wells's conception of the Martian red weed.[5]

In this article, Hamilton discusses both plant species that were 'purposefully introduced' and those that 'accidentally found their way here' (1893: 162). This distinction is echoed in *The War of the Worlds* in the narrator's statement that it remains unclear whether the Martians introduced the seeds of their planet's plants to Earth 'intentionally or accidentally' (Wells, [1898] 2009: 150). Hamilton stresses the haphazard means by which many species arrive in new environments. He relates an account of a European weed introduced to an Antarctic island by way of a spade brought from England with some earth still attached, recalls Darwin's observation of 'seeds being found in balls of clay attached to the feet of birds, and even to the elytra of beetles', and asserts that '[m]any aliens have arrived in [New South Wales] attached to the wool of sheep or the hair of other animals' (Hamilton, 1893: 162). He states as well that in many cases the means by which a plant first arrived in a new region was impossible to determine. In both his consideration of the means by which the Martian red weed was carried to Earth and his suggestion that no conclusive answer to this question is possible, Wells echoes contemporary scientific accounts of introduced species.

Further points of similarity between Hamilton's article and Wells's description of invasive alien vegetation are also evident. Hamilton notes that, of all the plants deliberately or accidentally introduced into an environment, 'a large number do not spread to any extent' and never become pests in human eyes (ibid.: 162). Wells makes a similar point by juxtaposing the wildly flourishing red weed with the less adaptable red creeper, 'quite a transitory growth' that never 'gained any footing in competition with terrestrial forms' (Wells, [1898] 2009: 150). Additionally, though Hamilton is primarily concerned with the impact of introduced species in Australia, he describes similar cases in New Zealand and America, demonstrating his awareness that the phenomenon of unruly introduced species was recurrent and was not confined to any single region of the globe. Wells can be read as extrapolating from this recurrent, world-wide pattern in his depiction of the same phenomenon on an extra-terrestrial scale. Hamilton's article is also suggestive in its recurrent reference to introduced species as 'aliens', a characterisation which, alongside Darwin's description of 'the invasion of the cardoon' (Hamilton, 1893: 162; Darwin, 1839: 139), suggests that Wells's tale of invasive alien plants can be read as an elaboration of contemporary scientific accounts of introduced species.

The long-term outcome that Hamilton predicts also anticipates elements of Wells's narrative. Consideration of a range of examples, particularly in America, suggested to Hamilton that while in many cases an introduced species 'at first completely beat the native, it is noteworthy that now the natives are holding their own' (Hamilton, 1893: 163). He quotes a New Zealand botanist, T. Kirk, as saying, 'At length a turning point is reached, the invaders lose a portion of their vigour, and become less encroaching, while the indigenous plants find the struggle less severe and gradually recover a portion of their lost ground' (Hamilton, 1893: 163). He also relates his own experience of reading 'doleful prophecies of the damage that might be expected when the Cape weed (*Cryptostemma calendulaceum*) became common [in Australia]' before watching '[i]t spread to a great extent in certain spots for a couple of years and then almost disappear' (ibid.: 163). These scenarios roughly parallel the fate of Wells's Martian red weed, wiped out by terrestrial bacteria against which it had no immunity. Hamilton's observations illustrate the contemporary awareness of a range of possible outcomes that might result from the introduction of a plant species to a new environment: complete failure to take hold; moderate success: rampant, uncontrollable spread; and temporary efflorescence followed by precipitous decline. Wells enacts a number of these possibilities through his presentation of the red creeper and red weed, suggesting his absorption of contemporary knowledge on the subject.

Beyond the general correspondence between Wells's representation of the Martian red weed and late nineteenth-century knowledge of introduced species, certain features of the red weed call up specific historical associations. The narrator of *The War of the Worlds* describes the red weed as 'cactus-shaped plants' with 'cactus-like branches' forming dense 'thicket[s]' that were sometimes 'knee-high', sometimes 'neck-high', and 'covered every scrap of unoccupied ground' in and around the decimated town of Sheen (Wells, [1898] 2009: 150, 167, 168, 203, 205). The narrator comments that he found the red weed 'broadcast throughout the country, and especially wherever there was a stream of water' (ibid.: 150). This description of the red weed recalls the prickly pear or *Opuntia*, a genus of plant in the cactus family that is native to the American continent, species of which were deliberately introduced to other hot and dry regions of the globe for ornamental and economic purposes. By the 1880s and 1890s, the prickly pear had become a serious pest in eastern Australia, southern Africa, India, and elsewhere, and accounts of its impact were beginning to be reported in the British scientific press.

The first reference to the prickly pear in *Nature* appears in the 15 March 1888 issue of the journal in an article by D. Morris entitled 'The Dispersion of Seeds and Plants'. Morris discusses the spread of seeds from one place to another by way of the digestive systems of both cattle and human beings, and he records that on the South Atlantic island of St. Helena it is not possible to use 'urban' manure in the neighbourhood of Jamestown, because 'if such manure was largely used, the land would become overrun with plants of the prickly pear, *Opuntia ficus-indica*, the fruit of which is largely consumed by the inhabitants' (Morris, 1888: 467). This allusion to the prickly pear relates in unexpected ways to Wells's representation of the Martians and their food sources. The advanced Martian invaders themselves have evolved in such a way that they no longer have digestive systems or the need to consume solid food; instead, they feed on the blood of the humanoid Martians that serve as their 'cattle'. So, whether deliberate or accidental, the introduction of the Martian plants to Earth likely relates to the use of the plants as fodder for these humanoid Martian 'cattle', a fact that echoes in an unsettling fashion Morris's discussion of the dispersion of plants by way of both humans and livestock.

Nature also drew attention to accounts of the prickly pear in other prominent scientific journals of the period, noting discussions of the prickly pear in the *Kew Bulletin* in 1888, 1890, and 1892 (Anon, 1888: 277; 1890: 573; 1892: 278). In the July 1888 issue of the *Kew Bulletin*, a memorandum from the Director of the Botanical Gardens at Cape Town, Professor MacOwan, records that in the districts of Somerset and Graaf-Reinet:

> The prickly pear has spread during the last 50 years or so as to become a serious difficulty. The courses of streams, and flats between their curvatures, have in many cases been completely over-run, and such places are generally abandoned in despair.
>
> *(MacOwan, 1888: 166)*

MacOwan describes the 'enlargement and increasing denseness' of 'noted thicket[s]' over the past ten years, and observing that individual landowners had attempted to 'rid themselves of the pest by simply throwing it, by waggon-loads, into the river', he comments, 'Of course, this simply passes the curse on to somebody unknown, living down-stream' (ibid.: 166). As will subsequently be shown, such details have parallels in Wells's novel. The same issue of the *Bulletin* also discusses the prickly pear in India, where the introduced plant had similarly come to cover 'immense tracts of country' (MacOwan, 1888: 170).

By the late 1880s, the prickly pear was perceived as a threat of sufficient serious-ness to warrant a concerted government response. In his 1888 article, MacOwan states that 'in this Colony and in Australia the thorny *Opuntia* has increased so much as to demand in some places legislative interference and Government expenditure for its extirpation' (MacOwan, 1888: 168). The September 1890 issue of the *Bulletin* reprints proposed Cape Colony legislation to combat the prickly pear that stresses the plant's potential to 'obtain a complete mastery' in a district and to spread from affected districts to adjoining areas (Wilmot, 1890: 186). It consequently calls for an Act 'to provide for the complete extirpation of the Prickly Pear' (ibid.: 187). The rhetoric of complete mastery and complete extirpation employed in this article frames the prickly pear problem as an all-or-nothing contest between human and plant that anticipates the stand-off between species depicted in *The War of the Worlds*.

At the same time, however, *Kew Bulletin* articles of the period display an awareness of the extent to which environmental context can determine the impact of a species. Noting that '[t]he spread of the Prickly Pear in South Africa has led to considerable interest being taken in the best means either to destroy the plants altogether, or to render them of some service in the rural economy of the country', the editor of the *Bulletin* published in the May/June 1892 issue an account of the prickly pear in Mexico, where it was a native plant (Fletcher, 1892: 144). The account, written by B. N. C. Fletcher, describes the prickly pear as a 'providence of nature', 'almost invaluable in hot, dry and specially sandy countries', and relates its use as food for humans, fodder for animals, and fuel (ibid.: 146, 145). This laudatory account of the prickly pear in Mexico illustrates a contemporary awareness of the way in which adaptations that suited a plant to its native habitat could render it a rapidly spreading pest species when introduced to a new environment. As will subsequently be demonstrated, Wells had a similarly clear conception of how the red weed had evolved to suit its Martian environment, and this informed his description of the plant's behaviour when introduced to Earth. Late nineteenth-century scientific periodicals display a nuanced understanding of the phenomenon of introduced species, and Wells's novel echoes this understand-ing in many particulars.

Up to this point, positing a connection between the Martian red weed and the prickly pear seems entirely plausible. However, this connection is complicated by the fact that subsequent descriptions of the Martian red weed in *The War of the Worlds* diverge markedly from the appearance and habits of the prickly pear. Leaving the town of Sheen, the narrator comes upon 'a brown sheet of flowing shallow water, where meadows used to be', which he later discovers to have been caused by

> the tropical exuberance of the red weed. Directly this extraordinary growth encountered water it straightway became gigantic and of unparalleled

fecundity. Its seeds were simply poured down into the water of the Wey and Thames, and its swiftly growing and Titanic water fronds speedily choked both those rivers.

(Wells, [1898] 2009: 170)

What has just happened? Suddenly this land-dwelling plant with cactus-like branches has become aquatic and possesses Titanic water fronds.

Upon first encountering this difficulty, I tried to reconcile these details with the facts of the prickly pear's dispersal. MacOwan's memorandum records the growth of the prickly pear in '[t]he courses of streams, and flats between their curvatures' (MacOwan, 1888: 166). It also notes the spread of the plant to new regions by way of flooding rivers and landowners' practice of disposing of uprooted plants by throwing them into waterways, scenarios that could be read as models for the spread of the Martian red weed by water. However, even if one entertains the possibility of a water-dwelling cactus, how does one account for the red weed's Titanic water fronds, which impede the narrator's feet as he attempts to wade through the shallow floods and which turn the river into 'a bubbly mass of red weed' (Wells, [1898] 2009: 171, 186)?

It is at this point that it becomes clear that no single historical example can wholly account for the Martian red weed, but this does not mean that historical models cease to be relevant. Following the example of Matthew Chew, I turned to the Canadian waterweed, *Elodea canadensis*, as an additional model for Wells's Martian red weed. The Canadian waterweed is a perennial aquatic herb native to North America that grows submerged in marshes, lakes, and streams and has a tendency to form large masses, which it did with considerable frequency following its accidental introduction to Britain in the early nineteenth century. Once introduced to Britain, it spread rapidly, choking waterways, before declining just as rapidly for no clearly discernible reason in the late nineteenth century. In an article in the 27 June 1872 issue of *Nature*, A. W. B. remarks on 'the suddenness with which the Canadian waterweed, *Elodea canadensis*, filled up all our canals and water-courses within a few years of its first introduction' (A. W. B., 1872: 164). In the 21 January 1886 issue of *Nature*, it is recorded that 'Mr Siddal writes on the American waterweed (*Anacharis alsinastrum*, Bab.) [a synonym for *Elodea canadensis*], its structure and habit, and adds some notes on its introduction into this country, the causes affecting its rapid spread at first, and present apparent diminution' (Anon, 1886: 279). By the late 1880s, the waterweed had already spread to its greatest extent and was recognised to be in retreat.

The Canadian waterweed thus provides a model for the Martian red weed's congestion of the Thames, which, as Wells's narrator recounts, caused the water of the river to pour 'in a broad and shallow stream across the meadows of Hampton and Twickenham', so that these suburbs were covered in 'sheets of the flooded river, red tinged with the weed' (Wells, [1898] 2009: 172). The red weed's subsequent fate also recalls the history of the waterweed in Britain. Wells's narrator reports that 'in the end the red weed succumbed almost as quickly as it had spread', due not to any measures taken by human beings against it but rather to the effects of 'a cankering disease' caused by terrestrial bacteria (ibid.: 171). This pattern of initial rapid spread followed by sharp decline, coupled with the lack of deliberate human involvement in the eradication of these plants, one actual, one fictional, is a significant point of similarity, for in *The War of the Worlds* Wells is keen to stress human beings' lack of agency in the face of even a plant pest.

Based on these multiple resemblances, the Martian red weed is a very strange plant indeed: an aquatic cactus, a terrestrial waterweed. Wells had an education in biology; he could be careless of details at times, but he would not have been unaware of the implausibility of his hybrid plant in an earthly context. The red weed is a deliberate fictional hybrid, drawing features from at least two identifiable species that had become nuisances when introduced to new environments.

The scientific rationale behind this hybrid plant is made clearer in an essay, 'The Things that Live on Mars', that Wells published in *Cosmopolitan Magazine* in March 1908, in which he explains his conception of Martian flora and fauna, 'developed in conformity with the very latest astronomical revelations' (Wells, [1908] 1993: 298). Recognising that Mars has less moisture than Earth, 'for we hardly ever see thick clouds there, and rain must be infrequent', Wells posits that '[s]ince the great danger for a plant in a dry air is desiccation, we may expect these Martian leaves to have thick cuticles, just as the cactus has' (ibid.: 300). Further speculating that 'moisture will come to the Martian plant mainly from . . . seasonal floods from the melting of the snow-caps', he reasons that the plants will be adapted to take advantage of the brief opportunity for growth offered by such seasonal flooding (ibid.: 300). Aware that in a low-gravity environment, thick stalks or trunks would not be needed to support the leaves and flowers of a plant, he argues that 'the typical Martian plant will probably be tall' and carry its cactus-like pads upon 'uplifting reedy stalks' (ibid.: 300). These 'fleshy, rather formless' yet 'slender, stalky' plants recall both the prickly pear and the waterweed and are at once contradictory and plausible creations (ibid.: 301). The adaptations that suit a Martian plant to its native environment give it an unlikely appearance by earthly standards and endow it with attributes and habits that cause it to become a rapidly spreading pest species when introduced to Earth.

However, if Wells's 1908 article elucidates the scientific rationale behind the appearance and behaviour of the red weed on both Mars and Earth, there is also an artistic justification for the composite nature of his alien plant, for the process of fictional hybridisation that Wells engaged in when creating the red weed also contributes in crucial ways to the overall impression left by this Martian organism. The Martian weed that Wells describes in his novel is an uncanny organism, an amalgam of familiar features that, taken together, form an unfamiliar species. Its otherworldly character is signalled not only by its red colouring but also by its contradictory appearance and habits, its cactus-like branches and Titanic water fronds, its spread over both land and water. How better to create an unsettling, destabilising sense of the unhomely or alien than by combining recognisable features and habits of widely different species in one impossible plant? In imagining a fictional hybrid of the prickly pear and the waterweed, Wells created a plant that could transcend existing taxonomic categories and breach habitat boundaries: a simultaneously terrestrial and aquatic plant that could spread rapidly and tumultuously through disparate environments and thus embody the threat of total invasion.

The spread of the Martian red weed echoes the advance of humanity's primary antagonist in the novel, the advanced Martian species, reinforcing the invasion theme through repetition, but the plant invasion also has its own distinctive significance. When, after days spent in hiding from the Martian invaders, Wells's narrator emerges into the open, he finds himself confronted with an 'accursed unearthly' landscape (Wells, [1898] 2009: 170). He states:

> I had not realised what had been happening to the world, had not anticipated this startling vision of unfamiliar things. I had expected to see Sheen in ruins – I found about me the landscape, weird and lurid, of another planet.
>
> *(ibid.: 169)*

The sight of a once-familiar landscape transformed by alien vegetation arouses in Wells's narrator a sense of defeat. He states:

> I felt the first inkling of a thing that presently grew quite clear in my mind, that oppressed me for many days, a sense of dethronement, a persuasion that I was no longer a master, but an animal among the animals, under the Martian heel. With us it would be as with them, to lurk and watch, to run and hide; the fear and empire of man had passed away.
>
> *(ibid.: 169–170)*

It is significant that it is not the sight of the advanced Martian species in their mechanical tripods that first inspires in the narrator this sense of dethronement, but rather the sight of a known landscape transformed almost overnight by the incursion of alien vegetation. Although, on a rational level, it is the technologically advanced Martian species that the narrator regards as having conquered humanity, it is the sense of powerlessness against an organism understood as less than, not more than, human, that acts as the trigger for this sense of dethronement.

It is the two-pronged nature of the Martian attack – an attack by both a species regarded as superior and a species assumed to be inferior – that succeeds in so wholly demoralising the narrator. The advance party of Martian invaders is in fact quite limited in number, made up of only a few individuals, and cannot maintain a continuous presence throughout the whole territory that it has seized; however, the uninterrupted coverage of the red weed creates a sense of pervasive occupation. The incursion of the red weed turns the landing of a handful of Martians into an experience of total invasion.

History and fiction inform and amplify each other via Wells's novel. Accounts of the spread of introduced species such as the prickly pear and the waterweed, the cardoon and the thistle, informed Wells's narrative of alien invasion, and Wells's fictional hybridisation of the features and habits of real plant species enabled his Martian red weed to become a multiple-habitat-spanning super-invasive. This process of amplification also took place on the level of language. Scientists such as Darwin and Hamilton had already adopted the terminology of invasion by aliens in their discussion of introduced species. However, by transposing these terms into a fictional tale in which the introduction of new species entailed the near-conquest of humanity by monstrous extra-terrestrials, Wells intensified the meanings that these terms carried.

These intensified meanings could then feed back into the non-fictional discourse surrounding introduced species. That the grafting of the fictional to the scientific could amplify the rhetoric surrounding introduced species in non-fiction texts is demonstrated by Wells's own popular science writing. *The Science of Life*, a compendium of modern biological knowledge published by H. G. Wells, Julian Huxley, and G. P. Wells in 1929–1930 and intended for a general audience, offered an account of the prickly pear in Australia. The authors describe the prickly pear as 'overrunning the country and ousting not only native plants, but man and his agricultural efforts as well' (Wells *et al*., 1931: 237). They recount:

> A few prickly-pears introduced into Eastern Australia as a botanical curiosity (and for a time propagated and spread by a kindly society who thought that cactuses in pots might brighten the homes of immigrants' wives) covered thousands of square miles in the course of a few years. At the height of its multiplication the prickly-pear was invading a new acre of Australian land every minute of the day, until, as Dr. Tillyard says: 'The vision arose of Eastern Australia becoming in a hundred years' time a vast desert of prickly-pear, with a few walled cities holding out against it.'
>
> *(ibid.: 598)*

The inclusion of this final dystopian image borrowed from Tillyard illustrates the way in which an amplified science-fictive perspective might gradually infuse the discussion of introduced species in non-fiction accounts, pre-conditioning responses to the phenomena described. That this process continues to this day is demonstrated by newspaper titles such as 'The War of the Weeds: The Invading Aliens Are Already Among Us', a discussion of invasive plant and animal species that takes the 2005 film adaptation of Wells's novel as its starting point (Burdick, 2005).

The rhetoric that surrounds introduced species has an impact on the decisions that we make regarding how to engage with these organisms. Sensationalist rhetoric inevitably intensifies our responses and, in so doing, potentially narrows our conception of the interactions that are possible. Wells is a useful case study for an examination of the amplification of rhetoric surrounding the representation of introduced species, for he was not only a writer of sensational science fiction narratives but also a populariser of science, a spokesman for science education, and a regular contributor to the most influential science journal of his time (and ours). Wells crossed and recrossed the boundaries between science and fiction, history and story, repeatedly over the course of his career, and it is worth thinking about what he carried with him on those crossings.

Notes

1   Kudzu was intentionally introduced to the United States in 1876 as a plant useful for controlling soil erosion, and its cultivation was encouraged throughout the first half of the twentieth century. In 1953, however, it was removed from the list of recommended cover plants; in 1970 it was declared a weed; and in 1997 it was added to the Federal Noxious Weed list (Britton *et al.*, 2002: 325).
2   I am grateful to Matthew Chew for drawing my attention to this reference.
3   In the 1859 edition of *On the Origin of Species*, Darwin does not name any individual plant species in the course of his discussion of introduced species; however, in the second edition of *Origin*, published in 1860, he makes specific reference to the cardoon and the thistle.
4   John S. Partington reproduces and analyses Wells's contributions to *Nature* and the journal's reviews of Wells's own work in *H. G. Wells in Nature, 1893–1946* (2008).
5   That Wells might consider accounts of European plants introduced to Australia in imagining the introduction of the Martian red weed to Earth is suggested by his reference in *The War of the Worlds* to the indigenous people of Tasmania, 'swept out of existence in a war of extermination waged by European immigrants, in the space of fifty years' (Wells, [1898] 2009: 3). Wells clearly regarded Britain's colonisation of a distant continent as an earthly model for his tale of interplanetary invasion.

References

Anon. 1886. Notes. *Nature*, 33, 276–279.
Anon. 1888. Notes. *Nature*, 38, 276–279.
Anon. 1890. Notes. *Nature*, 42, 573–576.
Anon. 1892. Notes. *Nature*, 46, 276–279.
Anon. 1897. Notes. *Nature*, 56, 13–17.
B., A. W. 1872. The Dispersion of Seeds by the Wind. *Nature*, 6, 164–165.
Britton, K. O., Orr, D. and Sun, J. 2002. Kudzu. In: Van Driesche, R., Lyon S., Blossey, B., Hoddle, M. and Reardon, R. *Biological Control of Invasive Plants in the Eastern United States*, Washington, DC, USDA Forest Service.
Burdick, A. 2005. War of the Weeds: The Invading Aliens Are Already Among Us. *Los Angeles Times*, 1 July.
Chew, M. K. 2009. The Monstering of Tamarisk: How Scientists Made a Plant into a Problem. *Journal of the History of Biology*, 42, 231–266.
Darwin, C. 1839. *Narrative of the Surveying Voyages of his Majesty's Ships Adventure and Beagle Between the Years 1826 and 1836, Describing Their Examination of the Southern Shores of South America, and the Beagle's Circumnavigation of the Globe, Journal and Remarks, 1832–1836*, London: Henry Colburn.
Darwin, C. 1860. *On the Origin of Species by Means of Natural Selection, Or the Preservation of Favoured Races in the Struggle for Life*, 2nd edn. London: John Murray.
Evans, D. H. 2001. Alien Corn: *The War of the Worlds*, *Independence Day*, and the Limits of the Global Imagination. *The Dalhousie Review*, 81, 7–23.
Fletcher, B. N. C. 1892. Prickly Pear in Mexico. *Bulletin of Miscellaneous Information (Royal Gardens, Kew)*, 65/66, 144–148.

Gregory, R. A. 1898. Science in Fiction. *Nature*, 57, 339–340.

Hamilton, A. G. 1893. The New Flora and the Old in Australia. *Nature*, 48, 161–163.

Haynes, R. D. 1980. *H. G. Wells, Discoverer of the Future: The Influence of Science on His Thought*, London: Macmillan.

MacOwan, P. 1888. Prickly Pear in South Africa. *Bulletin of Miscellaneous Information (Royal Gardens, Kew)*, 19, 165–173.

Morris, D. 1888. The Dispersion of Seeds and Plants. *Nature*, 37, 466–467.

Partington, J. S. 2008. *H. G. Wells in Nature, 1893–1946: A Reception Reader*, Frankfurt: Peter Lang.

Wells, H. G. [1898] 2009. *The War of the Worlds*, Cherrybrook, NSW: Horizon.

Wells, H. G. [1908] 1993. The Things that Live on Mars. In: Hughes, D. Y. and Geduld, H. M. (eds) *A Critical Edition of The War of the Worlds: H. G. Wells's Scientific Romance*, Bloomington, IN: Indiana University Press.

Wells, H. G., Huxley, J. and Wells, G. P. 1931. *The Science of Life*, London: Cassell.

Wilmot, A. 1890. Prickly Pear in South Africa. *Bulletin of Miscellaneous Information (Royal Gardens, Kew)*, 45, 186–188.

10

CANE TOADS

Animality and ecology in Mark Lewis's documentary films

Morgan Richards

There is a scene in *Cane Toads: The Conquest* (2010) that I found unexpectedly disturbing and sad. Volunteers from the Kimberley Toad Busters are shown filling plastic bags full of cane toads with CO_2, before unceremoniously dumping their lifeless bodies into a mass grave. A solitary cane toad is depicted watching on, a bystander to the attempted eradication of its own species, as a mournful soundtrack heightens the pathos. Despite being rendered environmentally aware by countless public campaigns and media stories detailing the impacts of cane toads on Australia's native animals and ecosystems, I could not escape an overwhelming feeling of sympathy for the toad.[1] I was aware of the aesthetic resonances of this scene, of its overt attempt to manipulate my emotions through the construction of a toad holocaust, but I still found it incredibly moving.[2] The animality of cane toads, or an awareness of the devastatingly violent price exacted on these animals for simply being out of place, an introduced or out-of-ecology species in a foreign landscape, kept disrupting the stability of my ecological convictions. This sense of ambiguity is key to writer-director Mark Lewis's approach as a filmmaker and his cinematic depiction of the complexity that surrounds cane toads in Australia.

Mark Lewis's documentary, *Cane Toads: An Unnatural History* (1987), and its sequel, *Cane Toads: The Conquest* (2010), spiral out from the deceptively simple narrative of how the toad, *Bufo marinus*, was introduced to North Queensland in the 1930s as a form of agricultural pest control. But these films are not simply an exercise in retelling this story and assessing the impacts, ecological and otherwise, of this invasive species. Instead Lewis radically shifts the framing of his films to focus on human–cane toad relationships. It is a stance that enables him to disrupt generic and species boundaries, unsettle conventional understandings of invasive species, and place the cane toad, quite literally, at the centre of his films. In a review of *Cane Toads: An Unnatural History*, writer and anthropologist Michael Taussig suggested that it deployed an 'extraordinarily effective method of sociological inquiry – namely the reading of human societal meanings into the animal kingdom' (1990: 1110). As Taussig makes clear, this is the opposite of anthropomorphism. Rather than projecting human characteristics onto toads, Lewis uses the intermediary of the toad as a means of illuminating the modern social world and refracting the complexity of the cane toad problem. As people from all walks of life offer their opinions on the toad, the self-conscious rendering of human–animal relationships allows a kind a transferral to take place, in which, as Taussig observes, 'the humans become somewhat like animals, and the toads become somewhat like humans' (ibid.: 1111). It turns out that the cane toad offers the perfect vehicle to blur various binaries including native/pest, domestic/wild and human/animal, in order to help us see the ecology of an invasive species in a multiplicity of forms.

This chapter examines Mark Lewis's cane toad films as an exemplary form of eco-documentary, with a view to understanding how they revise and overturn conventional forms of science and wildlife documentary. By examining the ways in which animals and humans are framed in Lewis's films, I wish to trace several cinematic and televisual trajectories that inform his practice as a non-fiction filmmaker, as well as looking at the broader impacts of these films. My central argument is that these films constitute an important model for eco-documentary, one that plays on the diverse, and sometimes conflicting, connections between humans and other animals. I am inspired, in part, by Jennifer Ladino's call for scholars of animal studies and ecocinema studies to examine films that depict 'ways of becoming with nonhuman animals', particularly films that blur generic conventions and invite discussion about the meanings and resonances of animality (2013: 144). My approach also draws on Anat Pick's 'creaturely poetics', in which she argues that the vulnerability that unites humans and non-humans can be used as 'a means of interrogating and expanding the possibilities of (non-human) subjectivity' (2011: 6).

My aim in this chapter is to open up new ways of thinking about invasive species, particularly cane toads, through the modalities of their popular representation in Mark Lewis's films. I begin with a brief exploration of how these films foster new ways of depicting human–animal relationships. Why did Mark Lewis choose to represent the cane toad in this particular way? I then consider the precise ways in which these films rework the differing scientific paradigms of conventional forms of science and wildlife documentary. By investigating Lewis's unique documentary style, it is possible to see how these films constitute an important form of eco-documentary. It is also possible to see the limits of this representational strategy. Finally, I look at the combination of animality and ecology in these films. Rather than seeking to master the problem of cane toads in Australia, what does it mean to respond to cane toads, as Lewis does in his films, from a variety of different perspectives?

Filming cane toads: from 'unnatural history' to 'conquest'

It is important to look at the techniques of representation in Mark Lewis's films as they are central to the way in which he politically and ethically reframes the invasion of cane toads, allowing new ecologies and nature–culture relations to emerge from the more conventional narrative of ecological disaster. *Cane Toads: An Unnatural History* (1987) is a 47-minute documentary, originally produced for television but later given a cinematic release.[3] It tells the story of how the cane toad, a species native to Central and South America, was introduced to Australia in 1935 in an effort to eradicate the scourge of cane grubs from Queensland's cane fields. The grubs of native French's cane beetles and greyback cane beetles (known commonly as witchetty grubs or cane grubs) were burrowing into the soil and attacking the roots of sugar cane, reducing crop yields by as much as 90 per cent. Under pressure to find a solution, scientists working at the Queensland Government's Bureau of Sugar Experiment Stations (BSES) seized upon a scientific study by entomologist Raquel Dexter, which appeared to link the introduction of cane toads to Puerto Rico with a corresponding drop in the population of white grubs that were decimating the nation's cane fields. However, the science behind Dexter's study was completely flawed (Turvey, 2010: 6–7)—a point that is left unexplored in the narrative of Lewis's film.[4] Nevertheless, through a visual montage that reconstructs the toad's journey and a series of interviews with scientists, government officials and farmers, the film explains how a batch of 102 cane toads was imported to Australia from Hawaii (another outpost of the sugar trade), before being bred and released at various sites around Gordonvale in north Queensland.

In spite of warnings from at least one prominent entomologist that the toad might become 'as great a pest as the rabbit or the cactus', a reference to the advance of prickly pear, the prevailing belief among BSES entomologists was that the toad would eat only cane beetles and other night-flying insects.[5] The film portrays the reasons behind the toad's importation as a mixture of commercial and government pressures that coalesced into the prospect of a successful biological solution, which was championed by many leading scientists. The reality, overlooked at the time by the assembled experts at the BSES, was that the toads and beetles rarely came into contact with one another. As Dr Glen Ingram, an expert on amphibians interviewed in the film reveals, 'The lifestyles of the cane beetles and the cane toads just didn't synchronise.' The beetles lived in the cane stalks and the toads lived on the ground, meaning that adult beetles were only vulnerable to the toads for the brief, hour-long interval in which they emerged from the soil, before flying off to forage in the trees.

Cane beetles were eventually controlled by the pesticide benzene hexachloride in 1945, which was marketed to cane farmers as Gammexane, until it was found to be carcinogenic and withdrawn from sale in 1987. But the toad proved to be remarkably adaptable to its new environment and quickly multiplied and spread. Its omnivorous eating habits and its possession of a highly poisonous bufotoxin, secreted from glands on either side of its head when it is attacked by predators, meant that it decimated populations of native Australian species—death adders, crocodiles, goannas, quolls (or native cats), kookaburras and ibises. Unlike indigenous prey, the toad's poison contains no unpleasant warning taste. In the absence of effective predators, and with a rapid breeding cycle, the toad's range quickly expanded. By the time Lewis set out to make his first film in the mid-1980s, cane toads occupied more than half the east coast of Australia, with an ever-increasing range stretching from northern New South Wales along the Queensland coast to the Northern Territory.

Dr Michael Archer, a mammalogist interviewed in the film, warns that the ecological impacts of cane toads will be catastrophic. 'What we are gradually going to see is one of these classic human disasters of a monoculture, gradually a single species replacing—and a single introduced species—replacing many many natural species.' He emphasises that the repercussions will be felt throughout entire ecosystems, and predicts that 'some of the dominant carnivores in those ecosystems are probably going to vanish'. Clutching eight toads, Dr Bill Freeland, a wildlife research officer in the Northern Territory, paints a disturbing ecological picture in which the Territory's wetlands will be transformed into 'a sea of little black tadpoles'. From there, he argues, the 'total conquest of Northern Australia is but a hop, step and a jump'. It is left to Freeland to deliver the film's final, deadpan words, 'At the moment we have absolutely no way of controlling the cane toad.'

This story of biological control gone wrong forms the backdrop of the film. But rather than simply focusing on the ecological impacts of cane toads, Lewis illuminates the synergies between cane toads and people. He focuses on cane toads as homophilic and synanthropic animals, emphasising their fondness for living in close proximity to humans and the ways in which they benefit from environments that have been altered by human habitation and industry, such as gardens, farms and roadsides. As Lewis argues: 'My films are as much about people as they are about animals. All of the films I've made are about animals that interact or are dependent or interdependent on humans for one thing or another' (Lewis, 2012).[6] Dismissing voice-of-God narration as an 'artificial device', he dispensed with it altogether in his film, using intertitles to bridge any gaps in the story (ibid.). Instead, he wove together interviews from an extraordinarily wide range of people—politicians, scientists, farmers, ecologists, children, environmentalists, government officials, retirees and local residents—who all offered their opinions on the toad.

Lewis was fascinated by how people had incorporated the toad into their everyday life, domesticating it, transforming it into a companion animal, and even utilising its poison to obtain an illicit high. He shows how cane toads and people have come to co-inhabit Queensland's cities and other human–culture environments, and how in the process the cane toad has become 'a highly ambiguous symbol of regional identity' (Thomas, 1991: 1119). Caricatures of cane toads regularly feature in political cartoons, most notably of the notoriously corrupt politician, Joh Bjelke-Petersen, who served as the Premier of Queensland from 1968 to 1987. Players in the Queensland rugby league team are affectionately known as 'cane toads'. But it is Lewis's focus on the diversity and complexity of individual human–cane toad relationships that forms the core of the film. This strategy effectively complicates the impulse to environmental education that structures most documentaries and news stories on the impacts of invasive species, which tend to focus on a singular narrative of invasion as catastrophe.

One of the most memorable characters is Monica, a young girl who is shown lovingly tickling her pet cane toad, the gigantic 'Dairy Queen'. Incidentally, a photograph of Monica and Dairy Queen was instrumental in obtaining funding for the film.[7] Another character, David Sondergard, a resident of Gordonvale who witnessed the introduction of cane toads, reveals how he likes to listen to the croaking of the toads as they mate in his backyard at night. These more companionable human–animal relationships are offset by depictions of far more malicious and violent interactions. Brent Vincent, a resident of Cairns, reaches ecstatic heights as he talks of the pleasurable 'pop' that can be achieved by running cane toads over in his Kombi van. Tip Byrne, a cane farmer, speaks of his desire to exterminate cane toads at every opportunity, exhibiting a tendency toward thinly veiled racism in his animosity for 'this creature who had been brought in'. At one point he even declares cane toads 'as big a menace as the German army'. Each has a story to tell about human–animal relations, and they do so in their own inimitable styles, direct to the camera and straight down the lens.

Lewis pioneered the use of a 'mirror box', a camera accessory he designed, which gives the impression that his interviewees are speaking directly to the audience. The sense of intimacy imparted through this technique was entirely new to the documentary genre. Nearly a decade would elapse before Errol Morris used a similar device in *Fast, Cheap & Out of Control* (1997), another film to focus on human–animal relationships.[8] Morris's 'interrotron' is renowned in histories of documentary film for its creation of a 'first person' narrative that renders the interviewee's words as private and intimate, somehow more truthful, even as it emphasises the constructed nature of his filmmaking process.[9] The fact that Lewis's mirror box is left out of many histories of documentary can be attributed to his position as an Australian documentary filmmaker with a predilection for making films about animals.[10] In other words, a filmmaker working in a marginal industry on the edges of a marginal genre. But it may also stem from the fact that he tended to frame his interviewees in mid-shots that placed them in the wider contexts of their homes and workplaces, adorned with the everyday clutter of their lives. By contrast, Morris preferred to use stylised close-ups of his interviewees, filmed in the stark confines of studio set-ups, which accentuated the direct gaze of his interviewees and the claustrophobic effects of his interrotron. Whatever the reason, Lewis's inventiveness as a filmmaker deserves greater recognition. His self-reflexive and animal-centric style of filmmaking can be understood as prefiguring the work of other documentary auteurs like Errol Morris and Werner Herzog, who later chose to focus on human–animal relationships.[11]

Lewis's use of experimental cinematography also extended to the depiction of cane toads. Throughout the production of the film, he worked with wildlife cinematographer Jim Frazier to film cane toads in a variety of different habitats.

Frazier's approach differed from conventional wildlife documentaries, which usually frame animals from the height of an average human or from the back of a truck, with the result that they tend to literally look down on their animal subjects.[12] By contrast, Frazier got down on the ground so that he was eye-to-eye with the cane toads he filmed. He also developed a set of lenses that allowed him to get incredibly close to cane toads, while still keeping the background of the image in sharp focus. This cinematic innovation, combining a wide breadth of field with incredibly deep focus, later gave rise to the Panavision/Frazier lens system, now widely used in wildlife cinematography and Hollywood cinema. With the aid of these lenses, Frazier was able to film cane toads against an array of different landscapes—farms, rainforests, wetlands, suburban backyards and cityscapes—depicting cane toads as part of broader environmental ecologies. He was also able to visually evoke the toad's journey through the use of roving point-of-view shots, in which the camera appeared to jump forward, mimicking the toad's advance into new territory.

In a wildlife television industry that was, even in the mid-1980s, already in thrall to charismatic mammals, Lewis elevated the toad, showing that its gleaming eyes, leathery flesh and pulsating throat could at times be equated with a kind of beauty. Although, in keeping with Lewis's distinctive documentary style, it is an unsettling and unnerving kind of beauty. His film also gave a rare voice to invasive species, which are almost completely disenfranchised by the market dynamics underlying the wildlife genre's restrictive focus on 'natural' environments, untouched by human culture and habitation. In his depiction of cane toads, Lewis charts a precarious course between anthropomorphism and searing cultural critique. Without ever actually slipping into anthropomorphism, or allowing the toad to stand as a straightforward metaphor for other things, he focuses on the cultural and political appropriation of the toad as a means of legitimating and furthering various social and political agendas from immigration to environmental politics and conservation. When talking about the multiplicity of issues that play across the figure of the cane toad in his film, Lewis remarked:

> Is it about the globalisation of invasive species and how we deal with this? Is it about bigotry? Is it about racism? Is it about immigration? Is it about refugees? Is it about all these different things that are popping up in our society? To some degree, many of these issues have got some sort of relevance or an analogy back to the toad. That was again, you know, the joy of the toad; it gave you a lot to work with.
>
> *(Lewis, 2012)*

Lewis's depiction of cane toads can be understood through what Cynthia Chris calls a 'zoomorphic framework, in which knowledge about animals is used to explain the human'. In this framework, as Chris argues, 'representations of animals articulate and reinforce new understandings of not only animal life but also human behaviour' (2006: x). This strategy of using knowledge about animals as a means of accessing or illuminating wider human social issues informs Lewis's practice as a filmmaker. At every stage, knowledge pertaining to the toad is utilised as a microcosm through which wider human social issues can be read.

A similar strategy is at the heart of *Cane Toads: The Conquest* (2010). As the first Australian feature film produced in 3D, it benefits from two decades of evolution in film technologies and the higher production values enabled by an exponentially larger budget. Beyond the spectacular aesthetics of HD cinematography and the visual poetics of 3D, it shares many similarities with the first film. A number of the original characters even appear again. Monica Krause, now grown up, reflects on the passing of Dairy Queen; Dr Glen Ingram returns to discuss the toad's anatomy and breeding cycle; Dr Mike Freeland riffs on the ways in which the toad has exceeded expectations, travelling faster and multiplying more rapidly than he predicted in the first film; while Tip Byrne, the archetypical angry farmer reappears, much older and wizened, to quip, 'They don't belong here at all—send them back to Hawaii—send them back over there to Barack Obama', updating his knowing racism and nationalism to a contemporary socio-political context. In the latest film, the 'deep north' of Queensland is transposed to the 'deep centre' and west of Australia, where a cast of Territorians are now battling or simply observing the toad's journey into Western Australia. Entrepreneurs have found other uses for the toad, processing its body as a resource for consumer products like handbags, Chinese medicines and fertiliser. A travelling 'Toad Show' of stuffed toads proved unpopular with Queenslanders, but a giant toad statue, erected in the Queensland town of Sarina, has become a tourist attraction.

One of the most distinctive features of Lewis's filmmaking practice is his use of humour. In *Cane Toads: An Unnatural History* he pioneered the approach of using humour as a means of subverting the wildlife genre and satirising the parade of dry experts that are usually wheeled out to offer their opinions in science documentaries. His films contain all of the usual ingredients of wildlife documentary, such as amplexus (or the toad's sexual habits), its reproductive cycle, its evolutionary adaptations, its omnivorous eating habits and its ecological interactions. But to these, Lewis added a range of stylistic effects—including devices borrowed from thriller, alien and horror films—which he uses to parody the familiar clichés of the wildlife genre. For example, in *Cane Toads: An Unnatural History* the toad's mating ritual is depicted through a reconstruction of a published account of a cane toad observed mating with a dead toad, a victim of road kill. In another scene, designed to send up the traditional predator-captures-its-prey sequence, a white mouse gingerly picks its way through a group of cane toads as a faux Hitchcockian soundtrack amplifies the tension. There is even a pastiche of Hitchcock's famous shower scene from *Psycho*, in which a man singing in the shower pulls back the curtain to reveal a large toad eyeing him from a distance. This scene, in particular, was intended as a comment on the sensationalised media coverage that cane toads usually receive, peppered with language designed to inspire anxiety and terror (Lewis, 2010: 23). Similar devices are deployed in *Cane Toads: The Conquest*. In particular, Lewis parodies the Disneyified vision of nature by depicting the act of cane toad amplexus as a thoroughly ironic 'romantic love vignette' set in an idyllic wetland (ibid.: 23).

This is a far cry from the scientific vision of the natural world offered in David Attenborough's wildlife documentaries. In many respects, Lewis's films provide the perfect antidote to Attenborough. 'If you're telling a story, you want people to see it', he revealed. 'I just found a lot of documentaries very dry and boring' (Lewis, 2012). In his own witty and campy way, he warns us about the artificial constraints of traditional wildlife and science documentaries. But beneath the humour there is far more going on.

Eco-documentary: subverting science and wildlife documentary

The disruptive power of Lewis's films emanate from their style. His approach of interweaving the often-contradictory perspectives of a wide range of interview subjects serves to destabilise the model of scientific expertise on which conventional wildlife and science documentaries depend, seeding doubt to the notion that science or ecology can provide the most salient perspectives on cane toads. But rather than eschewing this model altogether, his films self-reflexively engage with the idea of expertise, offering scientific facts while simultaneously playing with the notion of scientific and environmental expertise in ways that extend this knowledge into complex cultural and political contexts.

I suggest that Mark Lewis's distinctive documentary style provides an exemplary model for eco-documentary. These films do not represent a new genre of filmmaking. The success of Lewis's approach relies on the unique qualities of cane toads, as creatures that are simultaneously loved and reviled, and as homophilic animals that enjoy human contact and exhibit almost no fear in the presence of humans, allowing them to be filmed more easily. Lewis has successfully modified this approach to make films about rats and chickens, but beyond domesticated animals or synanthropic animals (wild animals that live in close proximity to humans) his approach has limited scope as a new format for wildlife documentary. What it does do is overturn conventional wildlife and science documentary formats, and in doing so it illustrates more refined ways of dealing with the complexities of environmental problems and the politics of invasive species.

Jennifer Ladino provides a framework for understanding the precise ways in which documentaries that focus on and complicate human–animal relationships display 'a companion species ethic', which she argues decentres the human-centric or speciesist perspective of many films and documentaries. She focuses on films, such as Morris's *Fast, Cheap & Out of Control* and Herzog's *Grizzly Man* (2005), outlining how they use experimental cinematography to collapse the distance between humans and other animals. For example, by allowing them to 'co-inhabit the cinematic space'; showing animals 'watching back'; using minimal dialogue or language to invoke the non-human world; and 'including zoomorphic footage and commentary', recalling Chris's conception of representations of animals that provide us with new understandings of animals and ourselves (Ladino, 2013: 131).

Lewis utilises many of these strategies in his cane toad films, but I would like to suggest that in extending a 'companion species ethic' to invasive species, his films provide a new model for understanding the complexities of the cane toad problem. Lewis's documentaries constantly flirt with and move between perspectives that seem to emphasise our shared animality with cane toads and perspectives that reinforce the (necessarily human-centric) ecological idea that cane toads need to be battled and controlled for the benefit of the environment. What saves his films from reinforcing the ideology that invasive species are always irredeemably bad—animals to be eradicated and killed with impunity—is his constant questioning of expertise and his construction of animal-eye-view shots.

First, then, to Lewis's strategy of undermining the status of scientific expertise as the foremost way of knowing animals. His decision to move away from the expertise of scientists to the practical knowledge of ordinary people was intentional. As Lewis argues:

> I wanted to focus on the people that knew the animal best, and they weren't necessarily scientists. Scientists generally specialise in animals and then they narrow that specialisation down to some sort of habit or element or attribute of the animal, whereas ordinary people are as good at observing animals as scientists and seem to know them just as well, if not sometimes better. So in other words the beekeepers, the people at home whose houses were invaded by cane toads, or the people that ran them over in their cars or who smoked them.
>
> *(Lewis, 2012)*

By calling on the expertise of ordinary people, Lewis was able to represent cane toads from a variety of different angles. In the process, he challenged the idea, implicit in the totalising narratives of many blue chip wildlife documentaries (which generally use voice-of-God narrations suffused with the knowledge of unseen scientific experts) that a single perspective could adequately explain the complexity of the issues surrounding cane toads.[13]

This sense of uncertainty about the superiority of scientific knowledge is also extended to the portrayal of scientists themselves. In contrast to the air of stiff-backed expertise that usually accompanies media portrayals of scientists, the biologists and other experts that Lewis interviews reveal their emotional attachments to animals (including cane toads) alongside their specialist knowledge. Perhaps the most humorous instance of this occurs in *Cane Toads: An Unnatural History* when Dr Glen Ingram mimics the cane toad's mating call, illustrating not just his intimate knowledge of cane toad behaviour but his clear admiration for the animal. He later reprised this scene, to similar effect, in the second cane toad film.

In another scene from the original film, mammalogist Dr Michael Archer mourns the loss of a western native cat, a research animal he also kept as a pet, which died after misadventure with a cane toad. Looking straight down the lens, he emphatically describes the sense of loss he felt at the death of this animal: 'Within twenty minutes that beautiful, unique animal, which I can only say I was totally in love with—this was something I was really wrapped up in—died in my arms in tetanic contractions.' Scenes like this serve a dual purpose. On the one hand, viewers are given a detailed explanation of precisely how the cane toad's bufotoxin works, acting on the cardiac muscles. By extension, the death of Archer's native cat also functions to emphasise the broader threats cane toads pose to native Australian mammals (later spelled out by Archer). And on the other hand, Lewis hints at his profound wariness of science as a practice inured to emotion, challenging the mode of scientific expertise that has so far dominated popular representations of ecological issues. With the aid of his 'mirror box', he also accentuates the blurring of official and public lives, allowing viewers to experience instances of what feels like direct eye contact with scientists and other officials, imparting a sense that these experts are sharing intimate knowledge with them. In this way, Lewis intensifies the subtle questioning of scientific expertise by emphasising the subjective perspectives and personal experiences of the scientists he interviews.[14]

The second strategy in Lewis's repertoire of techniques is his use of animal-eye-view shots to alternately reinforce and contradict the various ideologies espoused by his interviewees. Lewis argues that in both his films he tried to recreate the cane toad's perspective.

> We tried to tell much of the story from the cane toad's point of view, using exceptionally low camera angles—in effect, giving a voice to this animal that couldn't speak for itself yet was at the centre of so much controversy.
>
> *(Lewis, 2010: 25–26)*

This animal-eye-view characterises his approach to filmmaking, and is employed in two key ways. First, Lewis frames humans as if they are seen from a cane toad's perspective. This feature announces itself from the start of *Cane Toads: An Unnatural History*. The film begins with a number of interviewees, filmed from below in an obvious reference to Leni Riefenstahl's propaganda films but also mimicking the toad's viewpoint. The perspectives the interviewees put forward are jarring and wildly different, as these excerpts demonstrate: 'One male and one female in Darwin is more than sufficient to populate the entire top end of the Northern Territory'; 'The best thing is to get rid of them—get a big stick and hit 'em with it'; 'I couldn't do without them—they're friends.' The novelty of the camera's upward angle, combined with the humour implicit in seeing these characters as a cane toad might, looming above us, has the effect of creating a sense of uncertainty. The attentive viewer is invited to explore the political dimensions of each of the interviewee's arguments, exposing their ideologies, while also emphasising the myth of documentary realism.

Jim Frazier's innovative lenses, developed during the production of the original film, were instrumental in the creation of a second type of animal-eye-view, in which cane toads seem to look back at the viewer. At various points in the film cane toads are depicted sitting inertly, catching insects, looking directing back at the viewer, or simply watching on as the different stories the interviewees tell about them reveal the diverse ways in which people relate to cane toads. This strategy continues in *Cane Toads: The Conquest*, where the combination of spectacular 3D footage adds another dimension to the toad's story. In numerous shots, toads are framed against a shifting array of different backgrounds, huge skies, storm clouds, and vast, seemingly empty, vistas of arid scrublands and wetlands. Peter Hobbins argues that the depiction of cane toads against a variety of different landscapes, often emptied of people and other animals, has the impact of 'rhetorically naturalising them while glossing any local biotic accommodations or displacements' (2012: 86). In other words, it serves to obscure the ecological impacts of cane toads. This is certainly the case, though the wider narrative of the film enumerates the ecological impacts of cane toads. But these animal-eye-view shots also serve another purpose. By inviting viewers to focus, if only for a moment, on the shared animality of humans and cane toads, Lewis attempts to show us the vulnerability of the toad, not as a creature that has reached plague proportions, but as an animal that was introduced by humans and is now, all too easily, killed by them.

Through this 'creaturely poetics', to use Anat Pick's (2011) term, Lewis reminds us that the toad is powerless against ecological discourse. The threading of this second type of animal-eye-view shot throughout the films is used to suggest a critique of the cane toad's situation, adding an ethical dimension, which invites viewers to reflect on the cane toad's suffering. The shared animality of humans and cane toads is invoked most powerfully in scenes depicting the killing of cane toads. This is particularly evident in *Cane Toads: The Conquest*, where Lewis sets aside the ambiguity of the first film in favour of a more sympathetic portrayal. In the first film, Lewis chronicled the attempts of lone individuals to eradicate cane toads, occasionally veering into comedy as he showed Brent Vincent deliberately running toads over in his van (following Lewis's ethical stance against staging violence for the camera, the toads in the film were actually potatoes); or encouraged Dr Michael Archer to narrate the excruciating pain he endured from being hit in the eye with bufotoxin, jettisoned after he struck a toad with his geology pick. These acts of violence were constructed or reimagined for the camera. But in the 23 years since the first film, attempts to control the cane toad have blossomed into a full-scale industry, backed by scientists and politicians, and the violence Lewis portrays is more often than not real.

The film shows Northern Territory residents following local government advice to kill cane toads more humanely by freezing them and members of the Kimberley Toad Busters, a community group trying to stop the spread of cane toads, setting out on organised events to round up cane toads and gas them en masse. In examining the casual violence that is meted out to toads, Lewis never succumbs to the cinematics of slaughter. Each time a cane toad is killed, whether it is run over or gassed, this portrayal of real or constructed violence is undercut, either by the stories the interviewees tell or by the symbolic foreshadowing, achieved most powerfully in the toad holocaust scene, that violence against animals is closely related to violence against humans.

In scenes showing the activities of the Kimberley Toad Busters, for example, animal-eye-view shots of toads apparently watching on as members of their species are killed are used by Lewis to suggest that this is reminiscent of a toad holocaust. This might seem like cinematic hubris or blatant anthropomorphism, were it not for the overlapping narratives of the interviews that follow. In these interviews, Lewis alternately emphasises the absolute futility of killing adult cane toads and the desire, expressed by many different interviewees, to turn back the clock and reverse the devastating ecological impacts the toads have caused. Wildlife conservationist Dr Bill Freeland, for instance, debunks the idea that the actions of the Toad Busters are in any way effective: 'The invasion is going to go on, they are not going to stop it, they are actually wasting their time. They can't stop it, it's simply practically impossible'; while former Mayor of Katherine, Jim Forscutt, bemoans the lack of resources to battle toads: 'The federal government is giving six and half million dollars to save the whales, what about helping their own country [to] eradicate these blasted animals?'

In a later remark, Freeland reflects on the position of cane toads, inverting the idea that they are alien invaders: 'It's not the toad's fault that it came to Australia and caused the death of animals. It's our fault we brought the toads here. To them it's an alien country with alien animals and we're the cause.' Through this 'creaturely poetics', to use Anat Pick's term, Lewis invites us to explore the ethics of our own lives and the political dimensions of ecology. For Pick, humans and other animals are united by living bodies that are 'material, temporal and vulnerable' (2011: 5). This shared vulnerability opens up a means of moving beyond an anthropocentric perspective. Read through this prism, vulnerability can be understood as a key theme in Mark Lewis's cane toad films, extending out from the bodies of its human and cane toad subjects to encompass the broader vulnerabilities of scientific knowledge and forms of ecological control. By sticking closely to the cane toad, the film manages to document a multiplicity of relations and effects. Lewis demonstrates that it is possible to understand cane toads from both an animal-centric perspective (one that does not reduce them to a caricature of human behaviour, and takes note of their suffering) while still attending to broader political concerns about invasive species, ecosystems and their residents.

Using strategies such as the subtle undermining of scientific expertise and the creation of animal-eye-view shots, Lewis deliberately subverts the hallowed conventions of science and wildlife documentaries. By refusing to privilege any one perspective over another, letting the often-contradictory views of, for example, scientists and farmers sit side-by-side, he illuminates the complexity of the politics surrounding invasive species, raising profound questions about the political dimensions of both science and nature. His films present an alternate world, not often represented in science and wildlife documentaries, but much like our own, in which science merges with agriculture, local government and pet keeping. In other words, his films perform science as a product of culture.[15] At the same time, his vision of nature stretches far beyond the vision of pristine, untouched wilderness, which has become the stock-in-trade of conventional blue chip wildlife documentaries, to encompass a broader human–animal ecology.

Animality and ecology in Mark Lewis's cane toads films

Were it not for Mark Lewis's distinctive documentary style, several problematic belief systems would be at risk of being reconfirmed in the narratives of his films: namely, the privileging of humans as separate from natural environments and ecosystems, the superiority of native species over pests and other introduced animals, and the perception of cane toads solely as animals to be battled and controlled for the benefit of the environment. In both *Cane Toads: An Unnatural History* and *Cane Toads: The Conquest*, he succeeds in representing intersecting points of view—individual, agricultural, scientific, governmental, and even attempts to portray the cane toad's perspective through the construction of a range of animal-eye-view shots—extending the boundaries of wildlife and science documentaries. Never attempting to end up at 'truth', his films instead offer a complex mix of both animality and ecology.

The juxtaposition of different human–cane toads relationships in his films open up a possible spectrum of taxonomies, similar to those outlined by Adrian Franklin in his analysis of 'outside animals' in *Animal Nation*, from the harmless 'introduced' to the positive 'new native' and the emotive and thoroughly derisory 'noxious pest' (2006: 144). As illustrated in *Cane Toads: The Conquest*, the predicted devastation of native wildlife in the Northern Territory has failed to follow projected ecological models. New research by scientists like evolutionary biologist Rick Shine, who is featured in the film, indicates that 'ecosystems are much better at dealing with the impacts of cane toads than first indicated' (Shine, 2013). The work of Shine and other biologists demonstrates that cane toads have caused dramatic declines in populations of large predator species, such as Northern quolls, freshwater crocodiles and goannas, but after the initial invasion front populated by the largest and therefore most poisonous toads has past, predator numbers begin to climb again.[16] As Adrian Franklin argues: 'The cane toad demonstrates that animal ecologies are, despite the certainty of environmental predictions, prone to be surprising and changeable rather than fixed' (2006: 162).

Lewis's cane toad films are a perfect illustration of Franklin's argument for 'a cultural taxonomy of Australian animals'. As Franklin observes: 'The current wisdom on animals has established a confused and contradictory set of boundaries in which animals seem to suggest social and cultural flaws and fissures, anxieties and doubts in the national make-up rather than certainties and confidence' (2006: 153). It is this anxiety that Lewis mines in both his films. The slippery taxonomy of cane toads is something he constantly grapples with. Between competing depictions of cane toads as creatures to be deplored, loved and utilised for profit and the often hilarious one-liners played for laughs, there is a deeper message about how environmental issues are constructed out of an array of cultural and political concerns, even though they are often masked within a discourse of pure science. But, in the end, Lewis's greatest achievement as a filmmaker is to have made ecological or environmental documentaries that are eminently watchable. He doesn't preach to his audiences, instead, he offers a perspective that plays on the ambiguities and disjunctures of one of the biggest ecological disasters in Australia's history.

Acknowledgements

Thanks to Gay Hawkins, Graeme Turner and Elspeth Probyn for their insights and imaginative feedback. I would also like to thank the editors for their thoughtful questioning, encouragement and expert management of this project, especially Jodi Frawley, who got me thinking about cane toads.

Notes

1 In Australia, the problems posed by cane toads and other invasive species, as well as the various attempts to control or limit their impacts are perennially featured in the ABC's flagship science series *Catalyst* (2001–present) and its predecessor *Quantum* (1985–2001), which were both modelled on the BBC's *Horizon* (1964–present) series. Other ABC science series that have focused on the environmental impacts of cane toads and attempts to control them include *Toward 2000* (1981–1984) and *Beyond 2000* (1985–1995, 1999). *Landline* (1992–present), a series focusing on rural and agricultural issues, also occasionally features stories on cane toads and other invasive species, such as rabbits, foxes and camels.
2 This chapter is inspired by Gay Hawkins' chapter 'Plastic Bags' in *The Ethics of Waste: How We Relate to Rubbish* (2006: 21–44). In this chapter, Hawkins uses the example of the 'plastic bag scene' in *American Beauty* (1999) to explore how our ethical and political responses to waste and environmental catastrophe are sometimes shaped more by our affective and emotional responses to film and documentary than by official waste education campaigns. I use the example of the 'toad holocaust' scene in *Cane Toads: The Conquest* in a similar way. I am indebted to Gay for her insights on the aesthetic and ethical resonances of documentary films.
3 *Cane Toads: An Unnatural History* ranks as one of the most popular Australian documentaries. It held the record for the top-grossing Australian documentary at Australian box offices until 2008, when it was overtaken by *Bra Boys* (2007). It is currently ranked 87th in Screen Australia's 'Top 100 documentaries in Australia of all time', a list compiled using total reported gross Australian box office earnings. See http://www.screenaustralia.gov.au/research/statistics/wctopdocosalltime.aspx (accessed 9 May 2013).
4 Nigel Turvey, who is interviewed in *Cane Toads: The Conquest*, argues that Australia's problem with cane toads began with 'the myth of scientific proof' provided by Raquel Dexter's paper, 'The Food Habits of the Imported Toad *Bufo marinus* in the Sugar Cane Sections of Porto Rico' (1932). This paper analysed the stomach contents of 301 toads and concluded that since they contained 51 per cent of insects that were judged to be 'injurious to agriculture' that the toad's presence was causing the drop in white grub numbers. However, as Turvey outlines, this 'proved nothing at all about the dynamics of populations of toads and white grubs'. All it proved was what the toads had eaten for their last meals (Turvey, 2010: 6–7).
5 Walter Froggatt, a retired New South Wales government entomologist, issued one of the most prescient warnings about the cane toad in a paper published in the journal, *The Australian Naturalist*, in 1936. In this trenchant critique of the Queensland Government's release of cane toads, he warned: '[T]his giant toad, immune from enemies, omnivorous in its habits, and breeding all year round, may become as great a pest as the rabbit or cactus' (Turvey, 2010: 10).

6 Mark Lewis has built his career on films that focus on human–animal relationships, including: *The Wonderful World of Dogs* (1990), *Rat* (1998), *Animalicious* (1999), *The Natural History of the Chicken* (2000), and *Cane Toads: The Conquest* (2010). He has also produced two television series, *The Standards of Perfection* (2006), which focuses on show cats and show cattle, and *The Pursuit of Excellence* (2007), which includes an episode on ferrets and their owners.

7 A photograph of Monica and her pet cane toad Dairy Queen was widely published in Australian newspapers in the mid-1980s. Lewis was intrigued by this photo and he used it to 'sell' his programme idea to potential funders and distributors. *Cane Toads: An Unnatural History* was eventually funded by Film Australia and the BBC for a budget of around $160,000 Australian dollars. It was later broadcast nationally in Australia on the ABC in 1988 (Lewis, 2012).

8 Morris's *Fast, Cheap & Out of Control* (1997) documents the lives of four men whose work is reliant on non-human animals: George Mendonça, a gardener at a topiary garden in the US; Dave Hooper, a wild animal trainer; Ray Mendez, a naked mole rat expert; and Rodney Brooks, a robot scientist from MIT.

9 Morris's 'interrotron' is analysed in numerous analyses of the documentary genre. For a detailed account of his filmmaking practice, see *Three Documentary Filmmakers: Errol Morris, Ross McElwee, Jean Rouch* (Rothman: 2009), and Cunningham's interview with Errol Morris (2005).

10 The most notable exception to this is Brian Winston's account in *Claiming the Real* (1995), in which he cites Mark Lewis's *Cane Toads: An Unnatural History* as an important new direction for the future of documentary, which revives the tradition of 'biting social satire', seen in embryonic form in films such as *Land Without Bread* (1933). However, he makes no mention of Lewis's 'mirror box' (Winston, 1995: 255).

11 The disciplines of animal studies and ecocinema studies have spawned numerous analyses of the ways in which Herzog's *Grizzly Man* (2008) challenges traditional understandings of human–animal relationships. See, for example, Ladino (2013), Pick (2011) and Henry (2010). For a detailed analysis of how *Fast, Cheap & Out of Control* represents a 'companion species ethic', see Ladino (2013).

12 The wildlife docusoaps that came into vogue in the mid-1990s, spearheaded by the success of the BBC's *Big Cat Diary* (1996), provide one of the most notable exceptions to this. Docusoaps broke with the conventions of blue chip programmes, which had up to this point dominated the wildlife genre, by filming their animal subjects on hand-held digital cameras and often getting down on the same level as the animals they were featuring.

13 The term 'blue chip' has typically come to refer to programmes that depict spectacular visions of wild animals set in pristine and timeless wildernesses, devoid of any reference to human culture and politics, including environmental politics or conservation, which might date the programme or reduce its commercial appeal. As I have argued elsewhere, the marginalisation of environmental issues in the wildlife genre arises from the narrow scientific paradigm of natural history and the market dynamics of the wildlife television industry, which collude to ensure that controversial and politically challenging issues are suppressed (Richards, 2013).

14 Lewis's technique of placing his interviewees in specific contexts, for example, framing scientists in labs with white coats and politicians behind desks adorned with flags, added another layer of meaning. Rather than emphasising the power of their official positions, the clutter and regalia that surrounded them were used to demonstrate that their expertise was 'not necessarily as all-encompassing as they might like to think' (Lewis, 2010: 23).

15 Science documentaries, typified by the BBC's *Horizon* series (1964–present), conventionally use on-screen interviews with scientists and other experts to represent the contours of particular scientific debates. *Horizon*, in particular, has been praised for its ability to scrutinise science as a dynamic process, in which controversies erupt and wider social and cultural forces play a role (Silverstone, 1984; Jeffries, 2003; Darley, 2004). But as Andrew Darley observes, *Horizon* still presents viewers 'with assured and univocal

stories of discovery and progress' (2004: 232). By contrast, Lewis depicts the practice of science as radically inflected by culture. Through his depiction of the scientific and political blunders that led to the introduction of cane toads, as well as various unsuccessful attempts to control cane toads, he represents science as a fallible and thoroughly contested practice.

16 Research led by Professor Rick Shine has demonstrated that despite dire predictions of mass extinctions cane toads have had 'catastrophic impacts on large predators but not on other species'. Toads have, initially, devastated populations of Northern quolls, freshwater crocodiles, blue-tongued skinks, goannas such as the yellow-spotted monitor, and other varanid lizards and snakes in the Northern Territory. However, some species have benefited from the introduction of cane toads. Populations of certain species of native frogs and turtles, for example, have risen dramatically as a direct result of the decline in large predator numbers. Despite initial dramatic declines, populations of large predators have also stabilised as the invasion front has past. Shine's research has demonstrated that cane toads at the invasion front tend to be larger and more agile, and that once the invasion front has past, predators learn to evade the toads, getting sick rather than dying after attempting to eat smaller toads. Shine's team have also implemented conservation projects, such as 'predator training', in which Northern quolls are dosed with small amounts of cane toad bufotoxin to teach them to avoid the toads once they are released back into environments in which cane toads are present. Other species, such as native birds and rodents have proved to be less vulnerable to the toads (Shine, 2013).

References

Chris, C. 2006. *Watching Wildlife*, Minneapolis, MN: University of Minnesota Press.

Cunningham, M. 2005. *The Art of Documentary*, Berkeley, CA, New Riders.

Darley. A. 2004. Simulating Natural History: *Walking with Dinosaurs* as Hyper-Real Edutainment. *Science as Culture*, 12, 227–256.

Franklin, A. 2006. *Animal Nation: The True Story of Animals and Australia*, Sydney, University of New South Wales Press.

Hawkins, G. 2006. *The Ethics of Waste: How We Relate to Rubbish*, Lanham, MD, Rowman & Littlefield.

Henry, E. 2010. The Screaming of Silence: Constructions of Nature in Werner Herzog's *Grizzly Man*. In: Willoquet-Maricondi, P. (ed.) *Framing the World: Explorations in Ecocriticism and Film*, Charlottesville, VA, University of Virginia Press, pp. 177–186.

Hobbins, P. 2012. Review of *Cane Toads: An Unnatural History* and *Cane Toads: The Conquest*. *Historical Records of Australian Science*, 23, 86–87.

Jeffries, M. 2003. BBC Natural History Versus Science Paradigms. *Science as Culture*, 12, 527–545.

Ladino, J. 2013. Working with Animals: Regarding Companion Species in Documentary Film. In: Rust, S., Monani, S. and Cubitt, S. (eds) *Ecocinema Theory and Practice*, New York, Routledge, pp. 129–168.

Lewis, M. 2010. The Making—and Meaning—of *Cane Toads: The Conquest*. In: Weber, K. (ed.) *Cane Toads and Other Rogue Species*, New York, Participant Media, pp. 19–29.

Lewis, M. 2012. Interview by author, 22 November, digital sound recording.

Pick, A. 2011. *Creaturely Poetics: Animality and Vulnerability in Literature and Film*. New York, Columbia University Press.

Richards, M. 2013. Greening Wildlife Documentary. In: Lester, L. and Hutchins, B. (eds) *Environmental Conflict and the Media*, New York, Peter Lang, pp. 171–185.

Rothman, W. (ed.) 2009. *Three Documentary Filmmakers: Errol Morris, Ross McElwee, Jean Rouch*, Albany, NY, State University of New York Press.

Shine, R. 2013.Interview by author, 15 February, digital sound recording.

Silverstone, R. 1984. Narrative Strategies in Television Science: A Case Study. *Media, Culture & Society*, 6, 337–410.

Taussig. M. 1990. Review of *Cane Toads: An Unnatural History*. *American Anthropologist*, 92, 1110–1111.

Thomas. J. 1991. Review of *Cane Toads: An Unnatural History*. *American Historical Review*, 96, 1118–1120.

Turvey, N. 2010. The Toad's Tale: A True Fable of Science and Society. In: Weber, K. (ed.) *Cane Toads and Other Rogue Species*, New York, Participant Media, pp. 3–17.

Winston, B. 1995. *Claiming the Real: The Griersonian Documentary and Its Limitations*, London: British Film Institute.

11

WOLVOGS, PIGOONS AND CRAKERS

Invasion of the bodysplices in Margaret Atwood's *Oryx and Crake*

Peter Marks

'Invasion' usually suggests unwanted intrusion from outside a specific territory, be it a continent, a nation, a region, even a house. The title of this chapter references Don Siegel's 1956 film classic *Invasion of the Bodysnatchers*, which broadens the intrusive threat to planetary dimensions. In Siegel's film, the small Californian town of Santa Mira is invaded by extraterrestrials, who replace the local population with exact physical replicas of its citizens, but devoid of emotion or individuality. Because of when it was released, the film has sometimes been read as a Cold War allegory, the 'aliens' being equivalent to Soviet-inspired Communists. Ironically, others have interpreted it as signalling the rise of McCarthyism in the United States, while the author of the novel from which the film was adapted meant it as a warning against Eisenhower-era conformity (see LaValley, 1989). An intriguing aspect of the invasion is that the replicas arrive as spores from outer space, carried through Earth's atmosphere by the wind; they grow to maturity in bean-like 'pods'. Narratives depicting invasion from outer space date back at least to H. G. Wells's brilliant *The War of the Worlds* (1898), where the technologically superior Martians seem set, quickly and brutally, to take control of Earth. Again, there is a critical biological component, in this case, the terrestrial bacteria to which the Martians have no resistance, and which eventually destroy them. Where *Invasion of the Bodysnatchers* ends on a dramatically ambiguous note, the hero literally appealing to the film's audience to do something about the undetected invaders in their midst, in *The War of the Worlds* the natural environment rescues an otherwise defeated human race, destroying the would-be invaders.

Wells's generative genius also is traceable in a more recent novel, Margaret Atwood's 2004 speculative fiction *Oryx and Crake*. Atwood deals with an array of social, cultural, biological and ecological possibilities in her novel, including the gene-splicing of animals to produce hybrids such as 'wolvogs' (which map wolves on to dogs) and 'pigoons' (a mixture of pig and racoon) and the creation of a posthuman species, the Crakers. As we will see, Wells had considered similar hybrids over a century earlier in *The Island of Doctor Moreau* (1896). But where Wells confined experiments to an island controlled by the eponymous doctor, beyond which his creatures never really roam, Atwood's wolvogs and pigoons stalk malevolently around the post-apocalyptic world she fashions. This fits our sense of the contemporary Age of the Anthropocene, where human activity has massively influenced ecosystems around the globe, so that there is no 'outside' from which to invade. Everything is inside, including the invaders. The bodysplices in *Oryx and Crake* in a sense have been 'homemade' by humans—and in a traumatised environment where humans no longer dominate, some of these hybrid creatures are intelligent, vicious and hungry.

Bill McKibben's 1990 highly influential book on global warming, *The End of Nature*, is unforgiving of humans: 'we are no longer able to think of ourselves as a species tossed around by larger forces—now we are those larger forces. Hurricanes and thunderstorms and tornados become not acts of God but acts of man' (McKibben, 1990: xviii). It is a grim and slightly hyperbolic warning, for as recent tsunamis and earthquakes have shown so harrowingly, massive natural forces are sometimes just that—natural, immense and entirely uninfluenced by humans. That said, it is also apparent that human exploitation of the planet's resources has had a powerful, ongoing—and probably accelerating—impact on nature. Carolyn Merchant, in *Radical Ecology: The Search for a Livable World* (1992), has written of the 'confluence of technical and commercial orders' at work in this exploitative process (Merchant, 1992: 68). These orders impose what Graham Huggan and Helen Tiffen in *Postcolonial Ecocriticism: Literature, Animals, Environment* (2010) see as the 'technocratic view of nature', as something to be 'suitably retooled to match the latest global-corporate interests'. Atwood explores this retooling and its effects, intended and otherwise, in *Oryx and Crake*. The technocratic view produces a paradox, as Huggan and Tiffen note:

> Conventional environmentalist valorisations of nature run up against the obvious obstacle that much of what passes for the 'natural world' is a product of human activity and, once that truism is accepted, the 'nature' one is seeking to promote and protect isn't 'natural' in any autonomous sense.
>
> *(2010: 153)*

The unnatural scare quotes around 'nature' and its cognates suggest a breakdown of the definitional borders surrounding that term, to the point where the dividing line between the natural and the unnatural is constantly being redrawn and disputed, often as the result of technological advances generated by global-corporate interests. Again, we see this breakdown of borders (in several senses) in *Oryx and Crake*. That those interests do not take account fully of the social implications of activities that often transcend national or even continental borders dissolves the efficacy, if not indeed the reality, of borders, further undermining clear notions of what constitutes outside and inside.

Borders and God appear several times in this chapter. The world imagined by Atwood is 'global' in the sense that the environmental devastation unleashed upon it is planetary, but there are borders and spaces in which the main human character, who as a boy was called Jimmy but who in the environmentally desolate 'now' of the novel calls himself 'Snowman', can hide. Indeed, the novel opens with Snowman waking up in a tree, where he sleeps to protect himself against the predatory wolvogs and pigoons. He looks out on

> the eastern horizon [where] there's a greyish haze, lit now with a rosy, deadly glow . . . The offshore towers stand out in dark silhouette against it, rising improbably out of the pink and pale blue of the lagoon. The shrieks of birds that nest out there and the distant ocean grinding against the ersatz reefs of rusted car parts and jumbled rocks and assorted rubble sound almost like holiday traffic.
>
> *(Atwood, 2004: 5)*

Atwood conjures up a near-future world in which the natural and the unnatural— rosy, deadly glows; ersatz reefs; shrieking birds—play worryingly and ambiguously against each other. This account is, of course, only Snowman's interpretation, and it brings the human perspective in this troubled environment forcefully into view. 'Out of habit', we are then told, he looks at his watch, which has a blank face that prompts his thought: 'zero hour. It causes a jolt of terror to run through him, this absence of official time. Nobody nowhere knows what time it is' (ibid.: 5). Nobody nowhere because, as far as Snowman knows at the beginning of *Oryx and Crake*, he is the only living human, the last survivor before the onset of an entirely post-human world. Soon after, scanning the horizon where 'the sea is hot metal, the sky a bleached blue', he yells out 'Crake!', and when he receives no answer, adds, 'You did this!' (ibid.: 15). Crake, we learn, had been Jimmy's childhood friend, Glenn, who changed his name before setting out on a path to create an environmental apocalypse that indicates his delusion of God-like powers. In the true sense, though, Snowman is wrong, for though Crake certainly has been instrumental in wiping out seemingly all-but-one human (this is precisely Crake's plan), he cannot be blamed for the wolvogs and the pigoons and for much of the environmental degradation. Indeed, Crake's actions might be treated as a radical and diabolically successful response to the negative bodysplices and the destructively retooled world they represent. He might be interpreted not as the greatest sociopath in human history, but as the greatest eco-warrior in human history, who wants to bring that history to a close in order to save the planet *from* humans. The fact that wolvogs and pigoons continue to exist, however, threatens Crake's plan to repopulate the Earth with his own bodysplices, the Crakers, effectively calling into question his belief in his superhuman powers.

In order to make sense of what, even from these brief snippets, clearly is a complex and playfully provocative novel, let alone to put Snowman, Crake and the spliced creatures into context, we need to understand that there are two contexts detailed in alternating but interconnected narratives through *Oryx and Crake*. We begin in the 'future', somewhere around 2025, with a fragile Snowman struggling to survive in the world in part created by Crake. In the earlier narrative, which begins in the late 1990s, we meet a young boy called Jimmy (who, we come to recognise, becomes Snowman) and his new friend Glenn (who becomes Crake) growing up in a world marked by increasing environmental degradation: pyres of infected livestock; massive soil erosion; destructive storms and other undeniable evidence of catastrophic climate change. Glenn arrives, Jimmy recalls, 'in September or October, one of those months that used to be called *autumn*' (ibid.: 85), the italics underlining how weird the concept of seasons would become; when they graduate from high school, the highly intelligent Glenn would normally have gone to Harvard, had it not 'drowned' (ibid.: 211). Texas, we are informed, 'dried up and blew away' (ibid.: 295). And Jimmy watches a world in environmental freefall on 'the news: more plagues, more famines, more floods, more insect or microbe or small-mammal outbreaks, more droughts' (ibid.: 307). All this takes place in the early years of the twenty-first century, with Jimmy and Glenn as young, cynical and powerless onlookers.

Glenn, however, has a more sophisticated sense of the state of the planet and of the potential consequences of environmental catastrophe than does Jimmy:

> As a species we are in deep trouble, worse than anyone's saying. They're afraid to release the stats because people might just give up, but take it from me, we're running out of space-time. Demand for resources has exceeded supply for decades in marginal geopolitical areas, hence the famines and the droughts; but very soon, demand is going to exceed supply *for everyone*.
>
> *(ibid.: 356)*

The threat to Earth in *Oryx and Crake*, then, comes not from Wellsian invaders from another planet, but from Earth's seemingly most 'successful' species, *homo sapiens*. Glenn's plan (as the older Crake) to save the planet entails the simple if diabolical expedient of bringing the Age of the Anthropocene to an abrupt end by eradicating humans using a hyper-aphrodisiac called the BlyssPluss Pill that also generates plague-like death. Not that the humans who take the pill know this, of course. Indeed, Crake uses the unsuspecting Oryx to distribute the pill globally. Only later does she realise the connection, telling Jimmy, who is watching the plague's relentless impact: 'It was in the pills. It was in those pills I was giving away, the ones I was selling. It's all in the same cities. I went there' (ibid.: 390). The planet cleansed of humans will be gifted to a posthuman species Crake has developed and hubristically named 'The Crakers'. Crake's plan involves Snowman acting—unwittingly to begin with—as the Crakers' protector and guide, while they establish themselves as the environmentally-aware custodians of human-free world. That, at least, is Crake's plan, one that involves the rejection of the technologically reconfigured and damaged nature he grows up in, in favour, ironically, of the technologically retooled Crakers, who are the result of his own genetic experiments to build a species more amenable than humans to a sustainable planet. Crake has built his own death into the plan, and he projects that once Jimmy dies, having helped the Crakers settle into the posthuman environment, the new species will prosper. He has not, however, factored in herds of pigoons and packs of wolvogs; the unintended consequences of genetic engineering by others severely threaten his scheme.

The Crakers, then, from Crake's perspective, are the solution to the invasion of humans, whom he believes have ruined the planet. The Age of the Anthropocene, in this reading, is dangerously destructive to the planet as a whole, and must be terminated. And this in part distinguishes the benevolent Crakers from the wolvogs and pigoons, for the latter two creations are surviving manifestations of the earlier western philosophy that used the planet merely as a resource to fulfil human desires. It is this world that Crake and Snowman, as Glenn and Jimmy, grow up in, and get to know intimately through their parents, who work in various biological engineering communities financed by corporations. The pigoons and wolvogs represent aspects of what is, initially, a commercially successful enterprise geared towards providing the well-to-do inhabitants with endless supplies of food, while mollifying their fear of death with a vast and increasingly perverse array of life-extending or age-retarding treatments. Jimmy's father had been 'one of the foremost architects of the pigoon project', the goal of which

> was to grow an assortment of foolproof human tissue organs in a transgenic knockout pig host—organs that would transplant smoothly and avoid rejection, but would be able to fend off attacks by opportunistic microbes and viruses . . . A rapid-maturity gene was spliced in so the pigoon kidneys and livers and hearts would be ready sooner, and now they were perfecting a pigoon that could grow five or six kidneys at a time. Such a host could be reaped for its extra kidneys; then, rather than being destroyed, it could be kept living and grow more organs.
>
> *(ibid.: 27–28)*

Atwood in interviews repeatedly emphasises that much of *Oryx and Crake* was based on technology already available in 2004 when the book was published, and on experiments already being carried out. Certainly, arguments about the morality of such work had been well established before her novel, as in Donald and Ann Bruce's *Engineering Genesis: The Ethics of Genetic Engineering in Non-Human Species* (1998), which contains Ian Wilmut's case study, 'Xenotransplantation: Organ Transplants from Genetically Modified Pigs' (Bruce and Bruce, 1998: 63–66). And larger questions about the status of animals relative to humans go back centuries, the most influential modern text being Peter Singer's ground-breaking *Animal Liberation: A New Ethics for Our Treatment of Animals* (1976), where Singer argues against what he calls 'speciesism', discrimination on the basis of species. In Atwood's novel, speciesism abounds, other animals being merely raw material for human experimentation. As Traci Warkentin notes, *Oryx and Crake* provides 'a transitional narrative space for the discussion of current biotechnological philosophies and practices in Western society and where they might lead to in the not-so-distant future' (Warkentin, 2009: 152). Atwood's novel, then, acts as both an account of present-day practices and as a warning of the unforeseen or ignored dangers arising from such experiments.

Critically, a commercial dynamic underpins and finances the genetic work. The experiments carried out are undertaken in sealed communities or Compounds owned by companies whose names evoke the merging of biotechnology and business: OrganInc; HelthWyzer; RejoovenEsence. These corporations compete to satisfy a seemingly insatiable desire on the part of humans for food and eternal youth, the latter recognised by Crake as a response to '[g]rief in the face of inevitable death . . . The wish to stop time' (Atwood, 2004: 352). The desire to prolong human life might in some ways involve a conservative dimension, and in this sense the development of pigoons with multiple organs that can be transplanted into humans might involve a form of preservation. Yet the other driver towards endless consumption inevitably and rapidly exhausts the planet's resources. The suspicion increases that these forces are merging, so despite claims from OrganInc that the pigoons are not being used for bacon and sausages:

> as time went on and the coastal aquifers turned salty and the northern permafrost melted and the vast tundra bubbled with methane, and the drought in the midcontinental plains region went on and on, and the Asian steppes turned to sand dunes, and meat became harder to come by, some people had their doubts. Within OrganInc Farms itself it was noticeable how often bacon and ham sandwiches and pork pies turned up on the staff café menu. Andre's Bistro was the official name of the café, but the regulars called it Grunts. When Jimmy had lunch there with his father, as he did when his mother was feeling harried, the men and women at other tables would make jokes in bad taste.
>
> 'Pigoon pie again,' they would say.
>
> *(ibid.: 29)*

This especially upsets Jimmy 'because he thought of the pigoons as creatures very much like himself' (ibid.: 29). Given that their organs are being transplanted into humans, Jimmy's childhood concern raises perplexing questions about the ethics of xenotransplantation, as well as attendant debates regarding animal rights and species differentiation. Placing these considerations within the framework of rapid environmental degradation caused by humans creates an even more problematic, confronting and morally uncertain scenario, something the novel as a whole does not attempt to resolve. *Oryx and Crake* asks questions of readers given a peculiar though not impossible situation, but does not presume to answer them.

The breaking down of different sorts of defining lines, between spaces as well as species, remains an important aspect of the uncertainties Atwood consciously constructs, something she sees Wells also attempting in *The Island of Dr Moreau*. In

Wells's projection, Moreau carries out experiments on animals, so that, Atwood argues, '[b]oundaries between the normal levels of life dissolve: vegetable becomes animal, animal becomes quasi-human, human reverts to animal' (Atwood, 2005: xix). This breakdown has an impact on the species themselves and upon the accepted concepts and definitions of those who view the effects of these experiments. In *The Island of Doctor Moreau*, the narrator, Prendick, having escaped the hellish island and returned to 'civilisation', finds, Atwood plausibly argues, that

> [h]e lives in a state of queasy fear, inspired by his continued experience of dissolving boundaries: as the beasts of the island have at times appeared human, the human beings he encounters in England appear bestial.
>
> *(ibid.: xxvi)*

In her own novel something similar applies, for instance Jimmy's uneasy sense that the manufactured pigoons might be closer on a sliding scale to humans than are other animals, something that makes eating pigoons worryingly akin to cannibalism. Later, though, as Snowman, he faces a dangerous recalibration of power: with other humans dead, and the boundaries that had once contained the pigoons now broken, they treat him as food, and his waking hours are a constant battle to outsmart them and so survive a ghastly death.

The ability to outsmart pigoons is all the harder because of another piece of biogenetics in which Jimmy's father plays a leading role. Now working at NooSkins, he comes home drunk one night carrying a bottle of champagne to celebrate the success of 'the neuro-regeneration project. We now have genuine human cortex growing in a pigoon. Finally, after all those duds!' (ibid.: 66) For Jimmy's father, the project creates 'possibilities, for stroke victims, and . . .', but while he strives to find other positives, Jimmy's mother, herself once a committed microbiologist, but now disgruntled and 'harried', launches an attack against the 'moral cesspool' created by such work, and about its commercial basis—that the results will be sold

> at NooSkins prices . . . You hype your wares and take all their money and then they run out of cash, and it's no more treatment for them . . . Don't you remember about the way we used to talk, everything we wanted to do? Making life better for people . . . What you're doing—this pig brain thing. You're interfering with the building blocks of life. It's immoral. It's . . . sacrilegious.
>
> *(ibid.: 67)*

Whether or not we see the experiments on pigoons as sacrilegious, they do constitute a form of function creep, whereby the rationale and employment of a technology develop significantly beyond that for which it is originally created, leading to situations and problems undreamt of in its initial phase. As Jimmy's mother makes plain, the work on pigoons prompts major ethical questions. But years in the future, in the degraded world Snowman inhabits, the situation is reversed, the breakdown of barriers making the predicament existential rather than ethical: the pigoons stalk him as he goes in search of the food he needs to survive. When a herd of 30 of them trap him in a gatehouse, they are described as looking up at him through a window, seeing his head 'attached to what they know is a delicious meat pie just waiting to be opened up' (ibid.: 322). This chilling vision starkly inverts the model whereby humans plundered the Earth and its creatures; now, Snowman is on the menu, and the pigoons are untroubled by moral niceties.

The wolvogs, splices of wolves and dogs, exemplify another aspect of the world before the Crake-induced apocalypse—the need by corporations to protect the products, such as pigoons, that their scientists create, from theft by their competitors. Wolvogs also help protect the scientists themselves in a world growing increasingly competitive and anarchic. The Compounds are patrolled by the brutal and menacing CorpSeCorps, a security and surveillance force. CorpSeCorps has commissioned the high-level Watson-Crick Institute, where the adult Crake works in genetic engineering, to create a cross between a wolf and a dog with, as Crake informs Jimmy, a 'large pit-bull component'. These splices will be put into moats around the Compounds. Crake adds that the wolvogs are '[b]etter than an alarm system—no way of disarming these guys. And no way of making pals with them, not like real dogs' (ibid.: 250). Crake's work at Watson-Crick appears to show his acquiescence to the prevailing system, in which the brightest scientists are co-opted onto projects that, in the case of the wolvogs, bring in 'a lot of funding' (ibid.: 250). At its entrance, the institute has a statue of its mascot, a spoat, 'one of the first successful splices, done in Montreal at the turn of the century, goat spliced with a spider to produce high-tensile silk filaments in the milk. The main application nowadays was bulletproof vests' (ibid.: 242). The ChickieNob is another Watson-Crick triumph, a freakish creature with no head, which grows chicken parts on a massive scale for a ravenous market. 'This is horrible', says Jimmy out loud, the narrator adding: 'The thing was a nightmare. It was like an animal-protein tuber' (ibid.: 246). Horrible though it is, in a world of unrestrained consumption, such a creature is both necessity and a financial goldmine. Jimmy's visit to Watson-Crick, though, initiates a division between him and Crake. Seeing the wolvogs and ChickieNobs prompts Jimmy to think: 'Why is it that he feels some line has been crossed, some boundary transgressed?' (ibid.: 250). There is, in a cynical way, recognition of such unease, a biotechnologist explaining that 'they'd removed all the brain functions' (ibid.: 246) so that 'the animal-welfare freaks won't be able to say a word, because this thing feels no pain' (ibid.: 247). And while Jimmy 'couldn't see himself eating a ChickieNob. It would be like eating a large wart', he immediately qualifies his disgust by acknowledging that if the experiment succeeds 'maybe he wouldn't be able to tell the difference' (ibid.: 247). A broader sense of foreboding that incorporates a more physical notion of boundaries makes him ask Crake what would happen if the wolvogs got out.

'That would be a problem,' said Crake. 'But they won't get out. Nature is to zoos as God is to churches,' adding: 'Those walls and bars are there for a reason

. . . Not to keep us out, but to keep them in. Mankind needs barriers in both cases.'

When Jimmy asks who is 'them', Crake replies 'Nature and God' (ibid.: 250).

Crake has, however, been playing a double game. While insinuating himself into the highest levels of the genetic engineering industry, he is secretly planning to wipe out humanity using the very technological and scientific superiority that has maintained human dominion over the planet in the Age of the Anthropocene. Not only will his BlyssPlus Pill effectively eradicate *homo sapiens*, but he also has engineered a replacement species, the Crakers, to rule in their place. All this must be kept hidden, of course, and the Crakers are created at a clandestine research facility, Paradice, its name redolent of an earlier creation by an earlier deity. Crake does not make the Crakers in his image—given that he wants to wipe out his own species, that would be nonsensical—but he cannot resist giving them his name, and it is he alone who chooses their posthuman characteristics, including their colour: 'They're amazingly attractive, these children—each one naked, each one perfect, each one a different skin colour—chocolate, rose, tea, butter, cream, honey—but each with green eyes. Crakes's aesthetic' (ibid.: 10). Beyond the purely visual element, the Crakers have been engineered so that human traits such as competition, lust and envy are eradicated. Old age and fear of death, with the potential personal and social anxieties and disturbances they might trigger, have been eliminated, in that the Crakers are designed to drop dead at 30. And they are pre-adapted to the beleaguered environment of 2025, having 'a UV-resistant skin, a built-in insect repellant, an unprecedented ability to digest unrefined plant material. As for immunity from microbes, what had until now been done with drugs would now be innate' (ibid.: 366). Additionally, by altering 'the ancient primate brain', its 'destructive features' such as racism, the desire for hierarchy, territoriality and sexuality are controlled; sexuality is 'not a constant torment to them, not a cloud of turbulent hormones: they came into heat at regular intervals, as did most mammals other than man' (ibid.: 367). As Crake tells Jimmy:

> [The Crakers] were perfectly adjusted to their habitat, so they would never have to create houses or tools or weapons, or, for that matter, clothing. They would have no need to invent any harmful symbolisms, such as kingdoms, icons, gods or money. Best of all, they recycled their own excrement. By means of a brilliant splice, incorporating genetic material from . . .
>
> *(ibid.: 367)*

At this point Jimmy interrupts Crake's visionary explanation, but what is clear is the latter's unshakeable confidence that the Crakers represent an evolutionary response perfectly attuned to—and capable of rejuvenating—a denuded planet. What Jimmy does not realise at this point, and what Crake actively hides from him, is that this invasion of the bodysplices has been planned by Crake to supersede rather than to supplement humanity, and that Jimmy, reborn as Snowman, will play a critical role in the triumph of his project. Crake's plan depends on the invisibility of the Crakers to all but those within Paradice, and the concealment of his real plan from all others in the Paradice elect. Where Siegel and Wells had imagined invasions from without, from other planets, Crake's invasion comes from within.

Ultimately, despite Crake's megalomanic certainty, the collapse of the human civilisation that he brings about, a world without churches or God, is also one without zoos, so that the wolvogs and pigoons are free to roam, to menace and to kill. As Snowman realises, these malevolent splices easily have the power to wipe out Crake's idealised posthumans. And as well as those antagonistic specimens of that earlier, commercially-determined genetic engineering that Crake pretended to accept, but in fact detested, there are other forces in play that Crake has failed to envisage. Most obvious among these are humans besides Snowman who have, despite the catastrophe of the BlyssPlus plague, managed to survive in tiny numbers. Snowman had repeatedly wondered about this possibility, hoping that there might be

> Monks in desert hideaways, far from contagion; mountain goatherders who'd never mixed with the valley people; lost tribes in the jungles. Survivalists who'd tuned in early [to news of the global catastrophe], shot all comers, sealed themselves in their underground bunkers. Hillbillies; recluses, wandering lunatics, swathed in protective hallucinations. Bands of nomads, following the ancient ways.
>
> *(ibid.: 268)*

Most of these imagined survivors are less implicated in the excesses of human exploitation of the environment than the kind who would overindulge in ChickieNobs or slavishly desire the orgiastic sex activated by BlyssPlus. The novel in fact ends with Snowman discovering a small group of humans (who may or may not fit the list above) and not knowing whether or not to make contact with them, for there are unknowable consequences for the trusting, physically underpowered and dangerously innocent Crakers. Snowman's role as guardian stays his hand. And there is also the planet itself, which is beginning to recover from the ravages of both the Age of the Anthropocence and the consequences of Crake's attack upon the dominant species. When Snowman revisits one of the compounds, he realises on his way back that 'the botany is thrusting itself through every crack', and that soon this district will be 'impassable', a 'thick tangle of vegetation' (ibid.: 267–268). If impassable for him, then it is also potentially impassable for the much less physically robust Crakers. The very environment for which they were designed might imprison them. What these problems indicate is that Crake, despite his omniscient and omnipotent aspirations, has not foreseen all possibilities.

Crucially, his shortcomings as a would-be god also run to the Crakers themselves. Crake planned that Snowman, once he realised the full implications of what Crake had done, would protect the Crakers while they developed free from the drives and fear of death that had plagued humans, and began to live in harmony with a slowly recuperating planet. This plan depended on the genetic splicing and other forms of engineering making the Crakers not only a new environmentally-attuned species, but also the creators of a new form of social development. A central part of Crake's expectation was that, without such triggers as sexuality and the endless competition for resources, his improved creations would not follow the cultural path that had so influenced humans. Crake derides the art produced as a consequence of the prevailing human culture based on desire and fear, advising Jimmy:

> *Watch out for art*, Crake used to say. *As soon as they start doing art, we're in trouble.* Symbolic thinking of any kind would signal downfall, in Crake's view. Next they'd be inventing idols, and funerals, and grave goods, and the afterlife, and sin, and Linear B, and kings, and then slavery and war.
>
> *(ibid.: 430)*

But for all Crake's genetic manipulation, he admits that he cannot erase dreams and singing, which he recognises are hard-wired into the Crakers as much as into humans (ibid.: 419). These attributes, plus the questions the Crakers subsequently ask Snowman once Crake's plan has been activated, the stories Snowman tells them, their own processing of the information, and the symbolic signs and actions that they create, show the Crakers beginning to fashion rituals and belief systems that approximate human art and religion. So, for example, after Snowman returns from a brief journey back to Paradice, the Crakers tell him: 'We made a picture of you, to help us send our voices to you' (ibid.: 430). Given their reverential attitude to Snowman, we can tick art and idols off Crake's cultural list. Can Linear B—an archaic Greek script that combined syllables and pictographs, and that underpinned some of the intellectual achievements of Greek civilisation—be far behind? Or kings, slavery and war?

Such cultural developments, as repulsive to Crake as they might be, nevertheless require centuries, if not millennia. The question arises, whether, given wolvogs, pigoons, surviving humans and the environment itself, the Crakers stand any chance of surviving even a year. As regularly happens in Atwood's novels, she gives us no clues as to the likely outcome. Subsequent to *Oryx and Crake* she published another novel, *The Year of the Flood* (2009), which took the major planetary catastrophe of the earlier novel, retained some of its major characters such and Snowman, Crake and the Crakers, but interpreted them through a different lens, concentrating on another group of survivors, God's Gardeners. *The Year of the Flood* did little to resolve the questions provoked by *Oryx and Crake* (not that it was designed to do so). Atwood apparently plans a third instalment of this narrative, though it is unclear quite what this might entail. Not that we need these supplementary texts to interpret *Oryx and Crake*, and in fact their existence (or projected existence) reminds us of the creative diversity of interpretation itself. In projecting on from the end of the narrative of *Oryx and Crake*, without venturing too far beyond the boundaries of the novel itself, it seems that whatever immediate fate awaits the Crakers and the surviving humans, they will all have to take account of an environment more malevolent than that envisaged by Crake. Both groups will need to encounter and—if they are to survive—deal in some way with the numerous, predatory pigoons and wolvogs, those internal invaders now let loose among creatures ostensibly weaker than themselves. We might read this as something of a return to the prehistoric conditions of humans, forced by necessity to develop the physical, cultural, social and technological attributes and innovations that led over tens of millennia to survival and then to triumph, to the Age of the Anthropocene. Or we could instead project forward to a future we might have to endure sometime later this century, as the ecosystem we as a species have done much to damage reaches some tipping point. There probably will not be wolvogs, pigoons or Crakers in that potential environmental apocalypse. But *Oryx and Crake*'s dark fantasy requires us to consider that, as well as massive physical deprivations, we also could find ourselves threatened not only by the unwelcome machinations of other desperate humans, but also by the inhuman capacities and designs of insects, lions and eagles, among countless other creatures subject to the unforgiving law of the survival of the fittest.

References

Atwood, M. 2004. *Oryx and Crake*, Toronto, Seal Books.

Atwood, M. 2005. Introduction. In: Wells, H. G. *The Island of Doctor Moreau*, London, Penguin, pp. xiii–xxvii.

Bruce, D. and Bruce, A. 1998. *Engineering Genesis: The Ethics of Genetic Engineering in Non-Human Species*, London, Earthscan.

Huggan, G. and Tiffen, H. 2010. *Postcolonial Ecocriticism: Literature, Animals, Environment*, New York, Routledge.

Lavalley, A. 1989. *Invasion of the Bodysnatchers: Don Siegel, Director*, New Brunswick, NJ, Rutgers University Press.

McKibben, B. 1990. *The End of Nature*, London, Viking.

Merchant, C. 1992. *Radical Ecology: The Search for a Livable World*, New York, Routledge.

Singer, P. 1976. *Animal Liberation: A New Ethics for Our Treatment of Animals*, London, Jonathan Cape.

Warkentin, T. 2009. Dis/Integrating Animals: Ethical Dimension of the Genetic Engineering of Animals for Human Consumption. In: Gigliotti, C. (ed.) *Leonardo's Choice: Genetic Technologies and Animals*, Dordrecht, Springer Netherlands, pp. 151–171.

Wells, H. G. 1898. *The War of the Worlds*, London, William Heinemann.

Wilmut, I. 1998. Xenotransplantation: Organ Transplants from Genetically Modified Pigs. In: Bruce, A. and Bruce, D. (eds) *Engineering Genesis: The Ethics of Genetic Engineering in Non-Human Species*, London, Earthscan, pp. 63–66.

Film

Siegel, D. director, 1956. *Invasion of the Bodysnatchers*, Republic Pictures.

PART V

Unruly natives and exotics

12

INVASION ONTOLOGIES

Venom, visibility and the imagined histories of arthropods

Peter Hobbins

In 1988, historian Tom Griffiths planted a time bomb in Australia's natural history literature. Apparently unintended, this act of temporal terrorism raises intriguing questions about both what is 'natural' and what is 'historical' in ecology. Under the rubric of rethinking invasion ecologies, this chapter explores some of the epistemological and ontological questions raised by Griffiths' disturbing publication, particularly in relation to arthropods – the phylum of invertebrates that includes insects, crustaceans and arachnids.

Griffiths had presided over the publication of *The Life and Adventures of Edward Snell*, a nineteenth-century diary 'long regarded as one of the treasures of the State Library of Victoria's Australian Manuscripts Collection' (Snell, 1988: blurb). This lavish production reproduced Snell's extensive journal for the years 1849–1859, including his numerous sketches of colonial life, amateur ethnography and natural history. Among its lively drawings was specimen 35, a 'Venemous [*sic*] black spider with a red spot on his tail', which Snell encountered near Adelaide in November 1850 (Snell, 1988: 163; Figure 12.1). How he deduced that this petite specimen was 'venemous' is unclear: the only illness he detailed after the encounter was a hangover, and Snell made no allusion to either Aborigines or settlers suffering adverse effects from its bite. A recent immigrant from Britain, Snell likewise made no reference to related species elsewhere across the globe. Nevertheless, it is telling that he identified this spider as poisonous when he did not afford the same status to a large, hairy 'tarantula' encountered nearby earlier that same year (ibid.: 100).

One sense in which Griffiths' publication comprised a time bomb emerged in 1993, when Australian arachnologist Barbara York Main read Snell's account and declared it the first known record of the redback spider, now known as *Latrodectus hasselti*, anywhere on the Australian continent. Moreover, proposed Main, Snell's laconic description suggested that these spiders 'were well known (and possibly

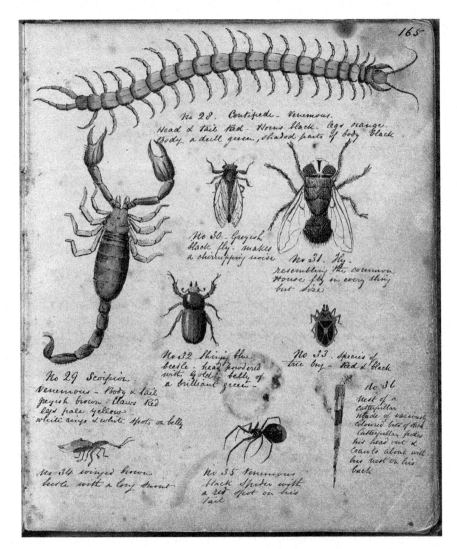

FIGURE 12.1 Page 165 of Edward Snell's journal features sketches of numerous Australian arthropods, including a formidable centipede, a scorpion and the 'venemous' spider now believed to represent a female *Latrodectus hasselti*.

Source: Reproduced, with permission, from the Pictures Collection, State Library of Victoria.

feared) as venomous' soon after Europeans occupied the Adelaide region in 1836 (Main, 1993: 3). This surmise was itself contestable, but of equal importance, it reopened a debate sparked by other Australian arachnologists. In 1987, Robert Raven and Julie Gallon had drawn attention to the absence of redbacks in forested areas, their ubiquity across urban sites, and the rather late formal description of the species in 1870, by which time over 200 other Australian spiders had entered the taxonomic literature. In adding that redbacks were first recorded at port cities – in a century thick-woven by its energetic maritime culture – Raven and Gallon ventured the mildly treasonous claim that 'Redback Spiders may not be native to Australia' (Raven and Gallon, 1987: 307).

This conjecture is the central problematic I wish to explore. But I am not seeking to determine the 'truth' of Raven's gambit; in fact, the question of the redback's indigenity is still, apparently, open. Likewise, my intention is not to narrate the competing claims that, since 1987, have characterised the debates between arachnologists, toxinologists, geneticists and ecologists over the origins and biological synonymy of *Latrodectus hasselti* (see, for instance, Low, 2001: 144–146; Griffiths *et al.*, 2005: 776–784; Vink *et al.*, 2008: 589–604). Rather, I want to step back – not just in time, but from the epistemological processes underpinning how we 'know' that a species is – or is not – autochthonous. What this chapter seeks to address is the ways in which we imagine histories for arthropods, conjuring up their putative existence before or beyond human contact. In short, their historical ontology. As I hope to make clear, the characteristics of *Latrodectus* spiders are particularly valuable for illustrating just how tenuous such ontologies are.

Amusing fictions and discursive invasions

The Europeans who arrived in the antipodean colonies by and large cared little about spiders. The nineteenth century commenced with Spanish soldier-naturalist Félix de Azara reporting that, when he induced enormous South American bird spiders (*Mygale avicularia*) to bite slaves, only a temporary fever ensued. At an 1848 meeting of the Linnean Society in London, British arachnologist John Blackwall likewise reported his 'Experiments on the Human Species' – himself – declaring claims that spiders might prove harmful to be mere 'amusing fictions' (Blackwall, 1848: 31–32). His dismissal was pointedly aimed at *Latrodectus* species and *Lycosa tarantula*. The bite of the latter, a wolf spider known in the Italian states as the 'tarantula', was popularly reputed to cause a potentially deadly lethargy cured only by frenzied dancing. As late as 1889, the influential American journal *Insect Life* reported a 'general incredulity among entomologists and arachnologists' that spider bites could prove fatal, citing many 'naturalists who have allowed themselves to be bitten without bad result' (Riley and Howard, 1889: 204).

Why, therefore, would a nineteenth-century settler in Van Diemen's Land, or the Swan River Colony, heed spiders? Moreover, why would any one spider prove more notable than another, apart from particularly remarkable characteristics such as size? In fact, prior to the 1860s, settlers' relatively rare references to Australian arachnids focused on ticks, scorpions or large, hairy 'tarantulas' like the example sketched by Snell – species now known generically as huntsman spiders (*Sparassidae* species). In 1890, Victorian amateur naturalist Charles Frost noted that:

> No small animal has more enemies, or fewer friends than the Voconia, or so-called 'Tarantula' – or perhaps more commonly 'Triantelope' – though,

why it received the name Tarantula I am at a loss to understand, unless it was from its supposed poisonous bite.

(Frost, 1890: 149–150)

While Frost was rare among his contemporaries in urging greater appreciation of spiders, suspicions that Australian 'triantelopes' were poisonous remained just that – suspicions. In the absence of Scriptural guidance as to their *natural* place or purpose, Europeans shunned spiders for their erratic movements or the 'guilt' implied by the fact that they scuttled away from human approach. Falling upon the heads of alarmed colonists also did little to endear them in the popular imagination (see Leigh, 1839: 131; Meredith, 1844: 147). Even where it was acknowledged that spiders might employ venom to capture their prey, concern that they might seriously poison humans remained a peripheral conjecture.

Thus, if a settler felt a sharp prick when lifting a piece of wood, donning a coat, or sitting on their privy, would they associate any ensuing unpleasantness with a spider's bite – even if they saw the presumed culprit? In other words, in the colonial antipodes, could a particular spider become visible via its venom?

Our answer comes from across the Tasman. Even before the Treaty of Waitangi was circulated among their diverse *iwi* in 1840, Māori people had informed the invading Pākehā of a deadly spider dwelling along the shoreline of New Zealand/ Aotearoa. In 1834, English missionary Richard Davis warned a colleague of these 'noxious vermin, called by the natives KATIPO. The katipo are very black, much like spiders, and have the property of the bug. When large, their bite produces inflammation, and sometimes death' (Coleman, 1865: 174). While subsequent settler panegyrics praised New Zealand's absence of snakes, by 1850 a combination of personal experience and Māori lore established the katipo as 'the only poisonous vermin in New Zealand', a narrative recapitulated with varying degrees of alarm over the ensuing century (Hochstetter, 1867: 440). This was not mere provincial credulity, nor unthinking ethnographic regurgitation. White settlers readily dismissed the potent *tapus* surrounding Aotearoa's autochthonous lizards, which Māori believed to possess malevolent spiritual powers (Wade, 1842: 178). In the case of katipos, however, Indigenous authority legitimated settler experience: at least 16 of the 25 bites reported up to 1870 – including all five of the reputed fatalities – were among Māori people (see Figure 12.2).

There was thus little doubt that venom was central to the cultural visibility of the katipo from well before European contact. The Māori lexicon contains few words denoting invertebrates. The generic phrase *Punga-were-were* encompasses almost all spiders, casting them alongside fish, lizards and insects as the loathsome offspring of Punga, son of the sea-god Tangaroa. Standing proud of this generic descriptor was the katipo, uniquely identifiable through its brown–black colour and red flash, its littoral habitat and especially its bite. Indeed, the katipo remains one of the few indigenous invertebrates to receive a specific Māori name (Vink *et al.*, 2008: 589). This semantic schism remained in place well beyond Waitangi. Throughout the century, 'spider' or 'insect' were used indiscriminately by colonists

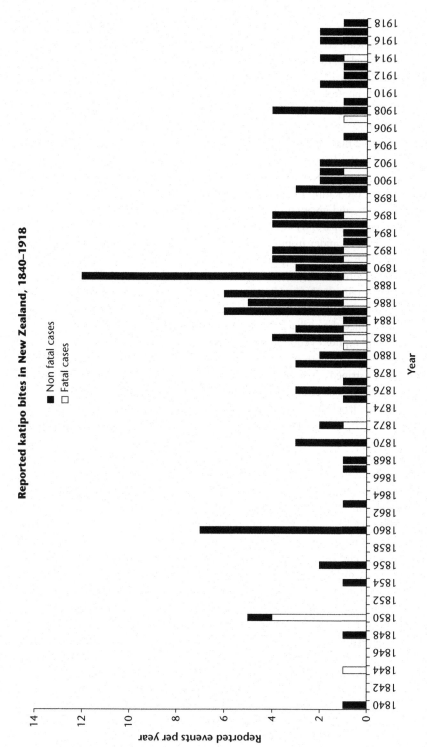

FIGURE 12.2 Numbers of katipo bites in Māori and Pākehā, as reported in colonial newspapers, scientific journals and other published sources, 1840–1918. Although many dates are approximations only, what is clear is that waves of concern over envenomation waxed and waned throughout the colonial period.

Source: Data collated and graphed by the author.

to describe the vast majority of arthropods, just as 'reptile' had long signified all manner of unclean 'creeping things that creep along the earth' aspersed by Leviticus 11:29. Nevertheless, by 1870, Pākehā usage of the 'katipo' clearly denoted the same spider described by Māori.

Furthermore, after this date a growing number of international medical and entomological texts cited the katipo as one of the world's few deadly spiders (Blyth, 1884: 447–448; Hirst, 1917: 10). As early as 1877, antipodean readers were outraged to find that the Belfast *Weekly News* was discouraging potential emigrants on account of the katipo, which was reputed to pose 'a great drawback to the enjoyment of life in New Zealand' (Anon, 1877b: 2). That same year, it was reported that 'the katipo of New Zealand seems to have made its appearance in North America, and great consternation has been the consequence'. Indeed, after specimens were apparently identified in Central Park, New York readers were informed that this spider inspired 'the dread of European immigrants and Maoris alike, and all observers unite in declaring that it is as deadly as the rattlesnake, or even as the nightmare of India – the cobra di capella' (Anon, 1877a: 3).

In the Australian colonies, however, there was almost no Aboriginal animus towards indigenous spiders or their bite, nor any specific word to describe the species now called the redback. Snell, after all, referred to his specimen only by its coloration and reputed venomousness, suggesting that in 1850 it had no common moniker. From the 1860s popular names began to proliferate, including the 'red-spotted', 'red-striped', 'red-back', 'black and red', 'jockey' or simply 'venomous' spider. Furthermore, after 1870, both local and international observers came increasingly to label the Australian *Latrodectus* simply a variety of the katipo (for instance, Karsch, 1884: 341). In this sense, the New Zealand species *discursively* invaded the Australian colonies.

Indeed, it was only in 1870 that both the katipo (*Latrodectus katipo*) and the redback – then labelled *Latrodectus scelio* – were formally described in the taxonomic literature (Powell, 1870: 58; Thorell, 1870: 369–374). Sadly the katipo type specimen has since been lost, but the archetypal redback still lurks inside London's Natural History Museum. Within biological discourse, one would typically state that 1870 thus represents the year in which the identities of both of these spiders became fixed. This is of course a simplistic assertion. Biologically, animal identities remain perpetually open to reorganisation and reclassification, from re-sketched cladograms to putative epigenomic maps. Culturally, attributes such as 'dangerousness' are constantly renegotiated in relation not merely to human experience, but to a matrix of animal interrelations (Hobbins, 2013).

But what of the period prior to 1870? Arthropods, especially specimens from the Anthropocene, are poorly served by the Australian fossil or archaeological record. Moreover, zooarchaeologists focus primarily on calcified artefacts reflecting human diets, such as fish bones and shell middens. Few pre-contact arthropod remains exist, given their chitinous exoskeletons and the ephemerality of soft tissues and bodily fluids like venom. While the reconstitution of biological molecules, including DNA and venom proteins, is technically feasible if samples can be secured, such studies cannot definitively reconstitute phenotypic expression (Weiner, 2010: 46–52, 56–59, 207–212). More generally, despite recent phenomenological interest in the (human) body, relatively few interpretive archaeologists have explored how their discipline could apprehend past biological characteristics or processes (Whitehouse, 2011: 227).

Outhouses and ontology

The year 1870, therefore, represents a rupture in the ontology of both katipos and redbacks. From that moment on, the encoding of these *Latrodectus* congeners within the scientific literature, coupled with the capture of type specimens, effectively ensured their *objective* existence as a species. I invoke here the process of 'objectivity' as theorised by philosopher of science Lorraine Daston. Daston has elaborated the operation of science as the conjunction of instruments and a community of trained observers who – by their consensual acknowledgement of what constitutes reliable data – permit the existence of entities, from cloud shapes to pathogenic bacteria. Drawing upon the seminal work of fellow philosophers Ludwik Fleck and Ian Hacking, Daston posits this process as historical ontology, as a means by which entities can be conceived to exist within a historical milieu (see Daston, 2000: 1–14; Daston and Galison, 2007: 1–53; Hacking, 2002: 1–26; Latour, 2000: 247–269).

But scientists are not the only community privileged to determine what can or cannot exist. Furthermore, their tools and techniques are not the only means by which 'real' entities come into being. As Thomas Anderson Stuart, Dean of the University of Sydney's Faculty of Medicine, observed in 1894: 'Records of cases of [*Latrodectus*] bite in man are neither numerous nor complete in the medical literature of Australasia, while the stories of the spiders are known to everyone' (Stuart, 1894: 10). Although there must be points of concordance that 'realise' and stabilise an entity within a historical moment, those points themselves can be subjective, quotidian or ephemeral. Amid the intersections of *Latrodectus* spiders with Indigenous and settler cultures, several such elements proved critical to the ontology of these arachnids.

Three specific characteristics of these spiders ensured that they became visible against a cultural background of general indifference to arthropods. Foremost was the subjective experience of envenomation. In 1870, New Zealand ornithologist Walter Buller observed of katipo bites that 'According to the natives, the common symptoms are an aching pain in the part bitten, which soon becomes much swollen and inflamed; then a copious sweat, and a feeling of intense languor and depression of spirits' (Buller, 1870: 32). Other reported symptoms included fever, migratory sweating and localised erection of body hairs. Nevertheless, both katipo and redback bites were primarily characterised nociceptively – as the embodied experience of pain. Until 1900, however, New Zealanders generally described this pain as more excruciating than accounts arising in the Australian colonies. Furthermore, while 15 per cent of katipo bites from 1844 to 1918 were reportedly fatal, the first deadly redback bite was not published until 1927 – albeit describing an event occurring 40 years earlier (Jackson, 1927: 525).

Yet envenomation is not simply a corollary of possessing fangs and a venom sac; it requires an embodied performance. Part of the credulity over the danger posed by *Latrodectus* – especially in Australia – was the spiders' reluctance to bite. They would often curl up and sham death when threatened. When in 1901 the Curator of Wellington's Colonial Museum allowed imported Australian redbacks to scurry across his hand, they did not bite. However, he reported, mere contact produced a local numbness he had similarly perceived when handling katipos (Anon, 1901). A further cause for credulity was the small size of the species and the correspondingly minute quantity of venom likely to be injected during a bite: 'if the Latrodectus stories are true', sniffed one American entomologist in 1889, 'we have a case in this creature of the most powerful poison known' (Anon, 1889: 347).

Both the geographical and anatomical location of bite sites also drew attention to a key behavioural difference between katipos and redbacks. Māori held that the katipo was a littoral species, dwelling in debris and foliage amid coastal sand dunes. This geographic locus is confirmed by my sample of 123 Māori and Pākehā bite reports from colonial newspapers. The bodily distribution of bites also reflected human beachside activities: many were sustained on the upper body by Māori who slept in the dunes, while exposed hands and feet proved similarly hazardous for Pākehā beach-goers. Furthermore, bites on the upper arms, thighs or torso often resulted after clothes were left on the sand and explored by wandering katipos.

Conversely, redbacks attracted particular opprobrium for the preponderance of bites on colonists' buttocks – and especially on men's genitalia. The first acknowledged redback bite, in 1863, prostrated a miner who, 'while sitting in a water closet felt himself stung in the lower part of the scrotum'. After hours of 'great agony', profuse sweating and laboured breathing, his penis had become 'much enlarged, the prepuce œdematous containing a large quantity of white serous fluid, the organ presenting the appearance in fact of one affected with virulent gonorrhœa' (Carr, 1863: 87). Much as it later proved the stuff of comedy, until well after the Australian colonies federated in 1901, the redback on the toilet seat was a truism almost anywhere that outside 'dunnies' were erected. While this peculiarity was shared with the American black widow (*Latrodectus mactans*), privies were far from the only places redbacks colonised. Nevertheless, their predilection for outhouses drew attention to the Australian species' fondness for sites of human settlement, often far from coastal regions.

The third *Latrodectus* quality of note was their distinctive coloration. It seems axiomatic that redback spiders should, indeed, sport a red mark on their backs. Yet this was never simply a question of phenotype; it was also one of Providence. Arising from seventeenth-century natural theology traditions, into the nineteenth century the 'doctrine of signatures' held that 'the most noxious animals have evident marks and characters, by which their dangerous properties are easily known' (Sturm, 1800: 90). Such biosemiotic shorthand had, for instance, already seen the Australian red-bellied black snake (*Pseudechis porphyriacus*) ostracised. In more recent times, similar attributes have demonised the blue-ringed octopus (*Hapalochlaena* species) or the white-tailed spider (*Lampona cylindrata*). Unlike the rapid Māori–Pākehā consensus over katipos, however, in Australia popular names proliferated, only stabilising on the 'redback' moniker in the 1920s.

Furthermore, even trans-Tasman exchange of specimens between museums and amateur naturalists failed to clearly determine the synonymy or otherwise of the Australian and New Zealand varieties. Which was the original, and which the doppelgänger? Which was native, and which the invader? For a time the redback was branded 'the Victorian katipo', but by 1900, British and American arachnologists concurred that 'the New South Wales spider is undoubtedly distinct from L. katipo' (Anon, 1893: 8; Cambridge, 1902: 253, 5, 8). In the Federation era, this distinction was not infrequently nationalist. '[W]hatever honour there may be in possessing it', wrote local naturalist Allan Wight in 1893, 'New Zealand has . . . a notoriously poisonous spider all to herself' (Anon, 1893: 8).

But what relevance do descriptive attributes – envenomation symptomatology, spatial distribution, or variations in colouring – have to do with rethinking invasion ecologies? In the colonial antipodes, these ostensibly inherent traits were not 'objective'. Rather, each was sensory, relational and impermanent. *Latrodectus* were not merely observed; they were *noticed*. As visiting German physician and evolutionary biologist Richard Semon remarked in 1899, 'This conspicuous spider seems to be everywhere' (Semon, 1899: 148). The identity of *Latrodectus* spiders was not a brute ontology of physicality – of material remains or reliably reproducible phenomena. Rather, a specific set of conjunctions fell into place, rendering these spiders visible against the biotic and discursive terrain of colonisation. This materialisation ultimately stabilised an ontology of spiders – and by extension all arthropods – as transhistorical *objects*.

Archaeology and arachnids

One reason why I am drawn to venomous animals is that they are so rarely configured as charismatic. Historically, few human observers have ascribed to snakes or spiders a meaningful intentionality or subjectivity. Unlike horses, dogs, or even imitative birds such as parrots, there is effectively nil intersubjective identification between humans and arthropods (Dirke, 2009: 161–163).

This was certainly true of nineteenth-century Europeans, whether their objects of scrutiny were domestic or colonial. In 1912, Belgian *littérateur* Maurice Maeterlinck wrote: 'We feel a certain earthly brotherhood . . . [with] other animals, the plants even, notwithstanding their dumb life and the great secrets which they cherish.' But for arthropods, he added, 'There is no question here of the human imagination. The insect does not belong to our world' (Maeterlinck, 1912: 9). Maeterlinck's assertion prefaced Jean-Henri Fabre's *The Life of the Spider*, the first widely read text to intimately describe the material world of arachnids. Yet despite years of close observation, Fabre never ventured to imagine the spider's interiority. Rather, like its external morphology or venom potency, he declared that a spider's behaviour was purely mechanistic, 'an inborn predisposition, inseparable from the animal' (Fabre, 1912: 75–76).

Although his works were less popular, Fabre's German contemporary – theoretical biologist Jakob von Uexküll – seemed to deny such materialistic determinism. His depictions of the sensory universe or *Umwelt* of animals – especially a 'detestable' arthropod, the tick – later proved highly influential in reconfiguring subjectivity in non-anthropocentric terms (Buchanan, 2008: 24–25). But despite depicting the *Umwelt* in relational terms – as environments simultaneously sensed and brought into being through each animal's interiority – Uexküll's arthropods remained mere vessels for morphological and behavioural archetypes. Denying spiders any autonomy, he asserted that their webs were successfully tailored to catch prey purely because 'an original program exists for both the fly and the spider' (Uexküll, 1982: 43).

Fabre and Uexküll were in any case outliers, thoughtful observers marking the end of a long century in which arachnids lurked below consideration. It might seem that this disregard for arthropod intentionality was amended after 1900. Consider the insect-creature's angst in Franz Kafka's *Metamorphosis*, or the altruistic spider central to E. B. White's *Charlotte's Web*. But such narratives operate via anthropomorphism, as recently satirised in Tim Ingold's imagined debate between Bruno Latour's ANT (Actor-Network Theory) and his own competing formulation of animate agency, SPIDER (Skilled Practice Involves Developmentally Embodied Responsiveness; Ingold, 2008: 209–215).

Less playful is the ethical mission of fostering trans-species identification that underpins much contemporary scholarship in posthumanism and animal studies. From Donna Haraway's dog to Jacques Derrida's cat, charismatic species remain privileged animal subjects (Derrida, 2002: 369–418; Haraway, 2003: 11–17). In recognising that animals also observe us, we do more than accept their alterity; we acknowledge their sentient otherness, co-constituting us both as moral agents. Indeed, critical ecological philosophers such as Val Plumwood require this conjoint acknowledgement of non-reciprocal moral relationships to break down anthropocentric subject–object dualisms (Plumwood, 2002: 178–184; Wolfe, 2010: 141–142).

As political projects, however, such schemas rely to a surprising degree on visual exchanges between charismatic, binocular mammals. Even if not anthropomorphised, a subject usually requires a face. Arthropods – with their gleaming exoskeletons and multiple eyes – remain animate but alien. *Les yeux sans visage*. They do not return our gaze.

My mission here is neither to grant spiders sapience, nor to extend to them the moral sensibilities that Marc Bekoff ascribes to non-domesticated mammals, from coyotes to red-necked wallabies (Bekoff, 2002: 120–132). Rather, it is to explicate this historical absence of intersubjective identification with arthropods that has constituted spiders – both morally and ontologically – as *objects* within ecological assemblages.

Of perhaps the greatest ontological import, arthropods – as objects – are fungible. It is a truism for invertebrates that each individual can stand in for, and hence be infinitely replaceable by, another of the same species. A robust legacy of Darwinism is its insistence on this gradual organic seriation: notwithstanding minor behavioural or morphological variations, the individual embodies the archetype (Wilkins, 2009: 227–234). Hence the biological fetishisation of type specimens. Indeed, the prevailing ontology of the arthropod is the prototypical specimen, an enduring object persisting from the remote past into the far future. A reverse transformation is effected in Kurt Vonnegut's *Slaughterhouse-Five*, in which the Tralfamadorians – extraterrestrials whose vision spans both time and space – see humans as quasi-arthropods, 'as great millipedes – "with babies' legs at one end and old people's legs at the other"' (Vonnegut, 1973: 62).

But how do we span the light years from Tralfamador to Tom Griffiths' time bomb?

I have gestured to 'assemblages', and it is to this concept that I now return to draw together the foregoing strands. There are two distinct genealogies of 'assemblage' that have recently begun to intersect in intriguing ways. The newer iteration, stemming from continental philosophy, has proved productive – if sometimes problematic – in crafting posthumanist accounts of animals or ecologies (for instance, Franklin, 2008). Extending poststructuralist attempts to decentre the subject, and with a sometimes under-acknowledged debt to Latour, such explorations incorporate the 'assemblages' narrated by Gilles Deleuze and Félix Guattari.

It is no coincidence that Deleuze and Guattari's constructs of anti-teleological multiplicity draw explicitly upon biological models: metastasis and becoming-animal. In eschewing Modernist tenets of autonomy, rationality and dominion, Deleuze and Guattari instead narrate assemblages as sites of productive intersection between perpetually unstable entities. Each entity – human or otherwise – simultaneously embodies both its archetypal iteration and its endless permutations: the wolf is becoming the pack, and vice versa.

Governed by no internal logos, or telos, these multiplicities are defined by their relations of exteriority – the points where they intersect with other multiplicities, contaminating them to spawn new assemblages: the wolf-man, the rhizome (Deleuze and Guattari, 2004: 256–287). Each assemblage thus reifies a historical singularity: its configurations are unrepeatable, and its ontologies unpredictable.

There is, however, an earlier usage of 'assemblage' that has – with some notable exceptions – been largely overlooked. The archaeological concept of assemblage is a fundamental disciplinary heuristic, ascribing meaning to a collection of artefacts. In archaeology, an 'assemblage' can denote an accumulation of disparate objects associated through contextual deposition or spatial relationships, embodying the material remains of a specific human culture. However, 'assemblage' also encompasses a find of similar objects within a bounded geographical site, representing variations on a type of artefact, or an industry. Within either definition, an archaeological assemblage is ultimately read as a single analytical unit.

What recent archaeological theorists have questioned is whether such assemblages are ever static. This is particularly so for scholars exploring the field known alternately as 'archaeology of the contemporary past' or 'archaeology in and of the present' (Harrison, 2011: 141–161). Laurent Olivier, for instance, analysed the Languedoc village of Oradour-sur-Glane, brutally liquidated by the Waffen-SS in June 1944. French attempts to memorialise the site, to preserve it 'as it was', led merely to 'the fabrication of hybrid constructions that belong as much to the present as the past' (Olivier, 2001: 183). Indeed, argued Olivier, any site, any assemblage that physically bears witness to human contact, is intrinsically archaeological, in that its elements represent 'a diversity of past temporalities', intersecting and contaminating each other (ibid.: 187).

Archaeologists typically address past cultures through artefacts: the enduring traces of human presence. This focus applies equally to zooarchaeology, archaeobotany and microarchaeology, which primarily examine biota to reconstruct their intersections with human cultures via their material remains. Artefacts, however, are ontologically defined not simply by their materiality, but by their manifestation before human observers (Hodder, 2001: 189). As Olivier asserts: 'Archaeological remains are inseparable from our present . . . it is they who need us if they are to exist' (Olivier, 2001: 180).

Yet if it is ontologically true that we see arthropods as transhistorical objects, rather than individuated subjects, then they too can be said to possess specific *archaeological* features within historical landscapes. Indeed, what drew me to archaeological theory was the discipline's historicising emphasis on interpreting conjunctions of space, place, event and object. Archaeology, as Gavin Lucas puts it, 'is a science of new entities, new assemblages' (Lucas, 2012: 265). Translating the twinned stabilising forces that hold assemblages together, he suggests that – in archaeological terms – the Deleuzian operation of coding equates to enchainment: semiotic operations linking the constituent objects. Concurrently, territorialisation equates to containment, defining a bounded site which perpetuates the assemblage (ibid.: 199–202). It is precisely these paired operations – semiotic linkages and biogeographical boundaries – which define ecologies both in scientific and historical discourse.

Furthermore, contends Lucas, historical materiality stands in an ambiguous ontological relationship with 'crude' physicality. Drawing on the work of phenomenological archaeologists, he argues against materiality as a static property of artefacts and their archaeological contexts. Rather, he posits a dynamic of materialisation, of *becoming* in the Deleuzian sense. For Lucas, ontology is perpetually (re)constituted both in sensory and relational terms, 'a process in which objects and people are made and unmade, in which they have no stable essences but are contextually and historically contingent' (ibid.: 166).

This is where Snell's specimen No. 35 crawls back into the picture. In nineteenth-century Australia – whether outside Adelaide in 1850, or in a Victorian privy in 1863 – redback spiders suddenly became visible. We read onto that visibility a rhetoric of ecological invasion: the spiders appeared where they had not been before. As arthropods leaving scant material remains, we will likely never know – in an 'objective' sense – whether or where they existed prior to colonisation. But our presumptive ontology of spiders as transhistorical objects intuits that if they had populated the pre-contact landscape, their features would not have escaped notice by earlier settlers, or Indigenous people. Yet animals are not archetypes, nor are traits such as coloration or venom potency invariant. Rather, these impermanent, sensory and relational characteristics – those operations which Lucas sees as fundamental to materialisation within a bounded site – are inherently anthropocentric.

Thus, it was European occupation that proved critical to materialising redback spiders within the assemblage of the colonial landscape. This was never merely a physical environment; it was also an imaginative one. Beyond the embodied experience of envenomation, the redback's visibility required a willingness to 'see' any spider as dangerous. Yet spiders played little role in Aboriginal or settler cosmologies until a species appeared whose reputation was, in turn, legitimated by Māori testimony. This operation of historical ontology was as tenuous culturally as it was biologically. After 1900, the katipo's fearsome reputation subsided markedly in New Zealand; by 1951, it was questioned whether any human had ever died from its bite (Hornabrook, 1951: 131–132). In Australia, panic over the redback has similarly waxed and waned. Rarely, however, has it been acknowledged that the ontology of its belonging was predicated on an invasion by outhouses.

References

Anon. 1877a. The katipo in the United States. *West Coast Times*, 13 June, p. 3.

Anon. 1877b. [untitled]. *Wanganui Chronicle*, 11 April, p. 2.

Anon. 1889. The spider-bite question. *Insect Life*, I, 347–349.

Anon. 1893. Entomological: the katipo spider. *Otago Witness*, 26 January, p. 8.

Anon. 1901. The bite of the katipo spider: notes by a medical man. *Evening Post*, 22 January, p. 7.

Bekoff, M. 2002. *Minding Animals: Awareness, Emotions, and Heart*, Oxford, Oxford University Press.

Blackwall, J. 1848. Experiments and observations on the poison of animals of the order Araneida. *Transactions of the Linnean Society of London*, XXI, 31–37.

Blyth, A. W. 1884. *Poisons: Their Effects and Detection. A Manual for the Use of Analytical Chemists and Experts. With an Introductory Essay on the Growth of Modern Toxicology*, London, Charles Griffin and Company.

Buchanan, B. 2008. *Onto-ethologies: The Animal Environments of Uexküll, Heidegger, Merleau-Ponty, and Deleuze*, New York, SUNY Press.

Buller, W. 1870. On the katipo, or venomous spider of New Zealand. *Transactions and Proceedings of the New Zealand Institute*, III, 29–34.

Cambridge, F. P. 1902. On the spiders of the genus *Latrodectus* Walckenaer. *Proceedings of the Zoological Society of London*, I, Part II, 247–261 and Plates XXVI–XXVII.

Carr, R. 1863. Two cases illustrating the dangerous effects resulting from the bites of venomous spiders. *Australian Medical Journal*, VIII, 87–88.

Coleman, J. N. 1865. *A Memoir of the Rev. Richard Davis, for Thirty-Nine Years a Missionary in New Zealand*, London, James Nisbet and Co.

Daston, L. 2000. The coming into being of scientific objects. In: Daston, L. (ed.), *Biographies of Scientific Objects*, Chicago, University of Chicago Press, pp. 1–14.

Daston, L. and Galison, P. 2007. *Objectivity*, New York, Zone Books.

Deleuze, G. and Guattari, F. 2004. *A Thousand Plateaus*, London, Continuum.

Derrida, J. 2002. The animal that therefore I am (more to follow). *Critical Inquiry*, 28, 369–418.

Dirke, K. 2009. Anthropologists in the world of insects. In: Holmberg, T. (ed.), *Investigating Human/Animal Relations in Science, Culture and Work*. Uppsala, Uppsala Universitet, pp. 154–163.

Fabre, J-H. 1912. *The Life of the Spider*, trans. A. Teixeire de Mattos, New York, Blue Ribbon Books.

Franklin, A. 2008. A choreography of fire: a posthumanist account of Australians and eucalypts. In: Pickering A. and Guzik, K. (eds) *The Mangle in Practice: Science, Society, and Becoming*, Durham, NC, Duke University Press, pp. 17–45.

Frost, C. 1890. Notes on the habits and senses of spiders – Part I. *Victorian Naturalist*, VI, 147–152.

Griffiths, J. W., Paterson, A. M., and Vink, C. J. 2005. Molecular insights into the biogeography and species status of New Zealand's endemic Latrodectus spider species; L. katipo and L. atritus (Araneae, Theridiidae). *Journal of Arachnology*, 33, 776–784.

Hacking, I. 2002. Historical ontology. *Historical Ontology*, Cambridge, MA, Harvard University Press, pp. 1–26.

Haraway, D. 2003. *The Companion Species Manifesto: Dogs, People, and Significant Otherness*, Chicago, Prickly Paradigm.

Harrison, R. 2011. Surface assemblages: towards an archaeology *in* and *of* the present. *Archaeological Dialogues*, 18, 141–161.

Hirst, S. 1917. *Species of Arachnida and Myriopoda (Scorpions, Spiders, Mites, Ticks and Centipedes) Injurious to Man*, London, British Museum (Natural History).

Hobbins, P. 2013. Imperial science or the republic of poison letters? Venomous animals, transnational exchange and colonial identities. In: Aldrich, R. and Mckenzie, K. (eds) *The Routledge History of Western Empires*, London, Routledge.

Hochstetter, F. von. 1867. *New Zealand: its Physical Geography, Geology and Natural History with Special Reference to the Results of Government Expeditions in the Provinces of Auckland and Nelson*, trans. Sauter, E., Stuttgart, J. G. Cotta.

Hodder, I. 2001. Epilogue. In: Buchli, V. and Lucas, G. (eds) *Archaeologies of the Contemporary Past*, London, Routledge, pp. 189–191.

Hornabrook, R. W. 1951. Studies in preventive hygiene from the Otago Medical School: the katipo spider. *New Zealand Medical Journal*, 50, 131–138.

Ingold, T. 2008. When ANT meets SPIDER: social theory for arthropods. In: Knappett, C. and Malafouris, L. (eds) *Material Agency: Towards a Non-Anthropocentric Approach*, New York, Springer, pp. 209–215.

Jackson, E. S. 1927. The red-backed spider bite. *Medical Journal of Australia*, I, 524–525.

Karsch, F. 1884. Die katipo-Spinne "laua-laua". *Berliner Entomologische Zeitschrift*, 28, 341–342.

Latour, B. 2000. On the partial existence of existing and nonexisting objects. In: Daston, L. (ed.) *Biographies of Scientific Objects*, Chicago, University of Chicago Press, pp. 247–269.

Leigh, W. H. 1839. *Reconnoitering Voyages and Travels, with Adventures in the New Colonies of South Australia; a Particular Description of the Town of Adelaide, and Kangaroo Island; and an Account of the Present State of Sydney and Parts Adjacent During the Years 1836, 1837, 1838*, London, Smith, Elder and Co.

Low, T. 2001. *The New Nature: Winners and Losers in Wild Australia*, Camberwell, Viking.

Lucas, G. 2012. *Understanding the Archaeological Record*, Cambridge, Cambridge University Press.

Maeterlinck, M. 1912. The insect's Homer. In: Fabre, J-H. (ed.) *The Life of the Spider*, New York, Blue Ribbon Books, pp. 7–35.

Main, B. Y. 1993. Redbacks may be dinky-di after all: an early record from South Australia. *Australian Arachnology*, 46, 3–4.

Meredith, L. A. 1844. *Notes and Sketches of New South Wales During a Residence in That Colony from 1839 to 1844*, London, John Murray.

Olivier, L. 2001. The archaeology of the contemporary past. In: Buchli, V. and Lucas, G. (eds) *Archaeologies of the Contemporary Past*, London, Routledge, pp. 175–188.

Plumwood, V. 2002. *Environmental Culture: The Ecological Crisis of Reason*, London, Routledge.

Powell, L. 1870. On *Latrodectus* (katipo), the poisonous spider of New Zealand. *Transactions and Proceedings of the New Zealand Institute*, III, 56–9 and Plate V.

Raven, R. and Gallon, J. 1987. The redback spider. In: Covacevich, J., Davie, P. and Pearn, J. (eds) *Toxic Plants and Animals: A Guide for Australia*, Brisbane, Queensland Museum, pp. 307–311.

Riley, C. V. and Howard, L. O. 1889. A contribution to the literature of fatal spider bites. *Insect Life*, I, 204–211.

Semon, R. 1899. *In the Australian Bush and on the Coast of the Coral Sea: Being the Experiences and Observations of a Naturalist in Australia, New Guinea and the Moluccas*, London, Macmillan and Co.

Snell, E. 1988. *The Life and Adventures of Edward Snell: The Illustrated Diary of an Artist, Engineer and Adventurer in the Australian Colonies 1849 to 1859*, T. Griffiths (ed.) Sydney, Angus & Robertson.

Stuart, T. P. A. 1894. Anniversary address. *Journal and Proceedings of the Royal Society of New South Wales*, XXVIII, 1–38.

Sturm, C. C. 1800. *Reflections for Every Day in the Year, on the Works of God, and of His Providence Throughout All Nature*, Edinburgh, D. Schaw and Co.

Thorell, T. 1870. Araneæ nonnullæ Novæ Hollandiæ, descriptæ. *Öfversigt af Kongl. vetenskaps-akademiens förhandlingar*, 27, 367–389.

Uexküll, J. von. 1982. The theory of meaning. *Semiotica*, 42, 25–87.

Vink, C. J., Sirvid, P. J., Malumbres-Olarte, J., Griffiths, J. W., Paquin, P. and Paterson, A. M. 2008. Species status and conservation issues of New Zealand's endemic Latrodectus spider species (Araneae: Theridiidae). *Invertebrate Systematics*, 22, 589–604.

Vonnegut, K., Jr. 1973. *Slaughterhouse-Five, or the Children's Crusade: A Duty-dance with Death*, Frogmore, Panther Books.

Wade, W. R. 1842. *A Journey in the Northern Island of New Zealand: Interspersed with Various Information Relative to the Country and People*, Hobart Town, George Rolwegan.

Weiner, S. 2010. *Microarchaeology: Beyond the Visible Archaeological Record*, Cambridge, Cambridge University Press.

Whitehouse, R. D. 2011. Cultural and biological approaches to the body in archaeology: can they be reconciled? In: Cochrane, E. E. and Gardner, A. (eds) *Evolutionary and Interpretive Archaeologies*, Walnut Creek, Left Coast Publishers, pp. 227–244.

Wilkins, J. S. 2009. *Species: A History of the Idea*, Berkeley, CA, University of California Press.

Wolfe, C. 2010. *What Is Posthumanism?* Minneapolis, MN, University of Minnesota Press.

13

NATURALISING AUSTRALIAN TREES IN SOUTH AFRICA

Climate, exotics and experimentation

Brett M. Bennett

Introduction

Planted Australian trees from the genera *Acacia, Casuarina, Eucalyptus* and *Hakea* have profoundly influenced South Africa's environment and economy over the past 130 years. White settlers in the Cape Colony imported and planted the first species of Acacia and Eucalyptus in the late 1820s. The economic value and total number of Australian trees raised in plantations sharply expanded in the 1880s and 1890s and continued growing throughout most of the twentieth century, especially in present-day KwaZulu–Natal and Mpumalanga. Today, there are approximately 500,000 hectares of Eucalyptus and 100,000 hectares of Acacia in South African plantations alone, without considering the substantial number of Acacia, Eucalyptus, and Hakea growing outside of commercial plantations (Republic of South Africa, Department of Agriculture, Forestry and Fisheries, 2008).

The question of why there are so many Australian trees in South Africa today has received attention from both scientists and historians. The vast majority of scientific studies examining Australian trees in South Africa have measured bio-logical and environmental factors to explain their dispersion patterns, though there is a small but growing movement among scientists to incorporate historical photography and environmental history. Many environmental historians also use scientific interpretations to explain the success of Australian trees. Beinart and Coates, for example, attribute the profusion of Australian species of Acacia and Eucalyptus to their biological ability to thrive in South Africa's environment: 'Fast growing (up to twenty feet in four years), stump sprouting, drought resistant and shade providing, these readily naturalized Antipodean migrants acculturated happily in relatively poor soil' (Beinart and Coates, 1995: 40). Numerous critiques of Crosby's 'ecological imperialism' thesis, which saw 'New World' (Australian and North and South American) floras and faunas as being weaker outside of their native habitats than their more dominant 'Old World' (African, Asian and

European) counterparts, use biological and environmental interpretations to counter-argue that there was an expansion of Australian Acacia and Eucalyptus in the Old World, including southern Africa (Crosby, 1986; Tomlinson, 1988: 89; Beinart and Middleton, 2004: 6, 10; Kull and Rangan, 2008; Radkau, 2008: 21–22, 159; Carruthers and Robin, 2010: 48–49).

Scholars more specifically examining the historical origins and development of state and private forestry in South Africa have not seriously analysed the diffusion or biological fecundity of Australian trees. Their efforts have focused on detailing the histories of individual species or genera, the intellectual origins of forestry laws and silvicultural models (Brown, 2001: 427–447; 2003; Barton, 2002: 98–103; Rajan, 2006; Bennett, 2010: 27–50) and the social history and effects of forestry policies on marginalised groups (Witt, 2002, 2005; Tropp, 2006; Pooley, 2010; Showers, 2010). Except for Barton's, which focuses on southern African-India connections, studies of forestry development have argued that South African foresters modelled their theories and practices directly upon German and French scientific precedents (Brown, 2001: Rajan, 2006: 78–79, 433; Showers, 2010: 305).

Brown, drawing upon the work of Rajan, uses a diffusionist model to explain the origins of forestry:

> An international model for forest management was thus being promoted – emanating from the European metropole and exported to the colonial periphery, where it was adapted to meet local conditions. The Cape was not a unique recipient of this transfer of technical knowledge.
>
> *(Brown, 2001: 433)*

This continental model gave professional foresters the power to create and manage state forests, advocated the strict separation of state forests from private property, and sought to convert existing diverse forests into homogeneous forests composed of a single species. But rather than examining the actual practices used by foresters, these studies trace South African forestry methods back to Europe by emphasising that many of the first foresters studied in Europe and referred to Germany and France frequently in their writings (Brown, 2003: 344–345; Rajan, 2006: 78–79. For a critique, see van Sittert, 2004: 311).

This chapter revises and reorients our understanding of the history of Australian trees in South Africa by challenging interpretations that argue that South African foresters directly applied European silvicultural practices and theories. It also adds a historical dimension that is lacking in many studies that emphasise biology and the environment as the critical factors in the naturalisation of Australian trees in South Africa. I argue that the widespread diffusion and naturalisation of Australian trees in South Africa resulted from a globally unique, ultimately successful state-sponsored research programme to select and then grow climatically suitable Australian genera and species in plantations.

Rather than embracing European forestry, South Africans actively rejected many European principles, such as the emphasis on managing large native forests; planted non-European flora; pursued the world's most extensive experiments with exotic species; and pioneered models to compare climates from around the world. German and French foresters, who cultivated and managed species native to Europe, could offer foresters in South Africa little practical help for selecting and growing exotic trees from outside of Europe. C. C. Robertson, a South African forester of the early twentieth century, noted: 'In the other branches of the science of Forestry, we can look to some other countries, and particularly to Germany . . . but the scientific naturalisation of exotic trees has so far received comparatively little attention in these countries' (Robertson, 1909: 219). Neither did studying in Europe necessarily predispose foresters to mimic European methods. The Cape forester David Ernest Hutchins, who studied forestry at the French national forestry school in Nancy, frequently argued that European practices would not work in the Cape Colony and helped found the first forestry school in South Africa. The methods South African foresters used to select species and create plantations remained highly controversial in European and British imperial forestry circles until the 1960s.

Australia's flora and Australian scientists played an important role in the origins of this experimental programme (Grove, 1995; Barton, 2002; Vandergeest and Peluso, 2006). Although Australian trees seemed to offer phenomenal results when grown in ideal conditions, the first seeds planted proved difficult to grow successfully. Millions of seeds planted out failed to grow; trees suffered various diseases, died prematurely and caused widespread disappointment for the people who planted them. In the 1890s, Cape foresters started working directly with Australian botanists to classify existing Australian species in the Cape and to select more climatically suited species. Hutchins and a growing cohort of South African foresters focused intensely on selecting and growing exotic trees. But because of the unique habits of the genus Eucalyptus, and the difficulties in matching exotic species to local sites, South African foresters had to build up their knowledge of Australian trees, quite literally, from the ground by establishing experimental trials across the country.

The first formal experiments with exotics began in the 1880s and 1890s when foresters in the Cape Colony started to select exotic species from climates similar to those in the Cape. Many species of Australian trees first entered South Africa in these experimental arboreta and plantations, where foresters tested them for their growth and technical properties. Within three decades, South African foresters had become world leaders in exotic plantation silviculture. In 1909, Robertson noted that 'probably more experimental planting of exotics has been carried out here than in any other part of the world' (Robertson, 1909: 219). After 1910, foresters working for the Union Department of Forestry created more rigorous experimental plantations and visited foreign countries to select suitable species to plant. Foresters in the first five decades of these experiments gradually learned what species grew best and how to plant and tend them. Throughout the twentieth century, state foresters and private industry drew upon this research to more confidently grow exotic trees in plantations.

Australian-southern African botanic transfers in the nineteenth century

The introduction of Australian trees should be situated within the larger history of biotic imports into southern Africa, beginning in 1652 with the first Dutch settlement in Cape Town. The settlers actively introduced exotic species of flora and fauna, creating new landscapes, environments and economies (Pooley, 2009: 19–20). White settlers in southern Africa planted exotic trees for several reasons, the most important of which was that few native forests existed. Today, closed indigenous forests cover less than 1 per cent of the land in South Africa, mostly scattered in small patches across the country (Mucina and Rutherford, 2006: 33).

The Dutch officially ceded the governance of the Cape Colony to Britain in 1814, a formal recognition of Britain's *de facto* rule of the Cape dating back to 1806. This official transfer encouraged new flows of people, plants and ideas between the Cape and Australia. The climates of southern Africa and Australia seemed similar to many of the people who travelled between the two continents. Many parts of the Cape Colony and settled littoral Australia have arid climates punctuated by drought and heat. Winter rain nourishes wheat-belts and vineyards in the southwestern Cape and the settled parts of South Australia and Western Australia. Many of the same types of seeds, weeds, plants and animals proliferated naturally in southern Africa and Australia (van Sittert, 2000; Beinart, 2003: 195–245, 266–303; Frawley, 2010).

White settlers in the Cape Colony began importing and planting Australian Acacia and Eucalyptus starting in the late 1820s. They planted exotic trees for timber, for shelter from the sun and wind, and because they believed trees increased rainfall in dry environments (Grove, 1987: 21–39; Barton, 2002: 98–105). Amateur naturalists and travellers idiosyncratically directed the earliest biotic exchanges before the creation of state and private botanic gardens in the middle of the nineteenth century. Popular historical accounts suggest that *Eucalyptus globulus* (blue gum) was introduced into the Cape in 1828 when Sir Galbraith Lowry Cole, the new governor, brought the species with him from Mauritius (Noble, 1886: 150). The first known introduction of Australian *Acacia longifolia* came in 1827 when James Bowie, a plant collector for Kew Gardens, arrived in the Cape with seeds from England (Shaughnessy, 1980: 104–105).

The popularity of Australian trees soared among Cape colonists in the 1860s after botanical enthusiasts in Australia and France peddled grandiose claims about their properties, especially those of the genus Eucalyptus. The Australian botanist, Ferdinand von Mueller, boasted that the timber of *Eucalyptus globulus* rivalled the world's most valuable timbers. Mueller and other botanists also argued that eucalypts helped to cure malaria and other tropical diseases both by draining swamps because of their vigorous growth and through the secretion of their scented, powerful oils, which subscribers to the miasmic theory of disease believed would kill malaria (Bennett, 2010: 30–32).

The mid-Victorian belief that Australia and southern Africa's locations in the southern hemisphere made them geographically and botanically related helped settlers to naturalise Australian trees. The botanist Joseph Hooker, who noted similarities among the floras on different continents and islands across the entire southern hemisphere, hypothesised that Australia and southern Africa might have been bridged together as part of a large ancient southern continent (Hooker, 1853, 1860). The popular and prolific naturalist Alfred Russel Wallace argued that the southern hemisphere was characterised by 'detached areas, in which rich floras have developed . . . but [are] comparatively impotent and inferior beyond their own domain' (Wallace, 1880: 495). The only exception to this rule was Australia's forest flora, which grew in the southern hemisphere outside of its original geographic range (ibid.: 496). Scientists in the Cape Colony expressed similar beliefs. John Croumbie Brown viewed the dominance of Australian trees in the Cape in the 1860s as a sign of their evolutionary superiority (Grove, 1989: 184, from Beinart, 2003: 41). At least one Cape Colony forester, Hutchins, noted the dominance of the Australian flora over the Cape's when arguing for the importation of Australian trees into southern Africa (Hutchins, 1905: 18–19).

The creation of new scientific institutions in the second half of the nineteenth century increased the intensity of biotic exchange. Melbourne and Cape Town both established botanic gardens in the 1850s (McCracken, 1997; Drayton, 2000). Gardens opened in Adelaide, Durban, Pietermaritzburg, King William's Town, Graaff-Reinet, Perth and Grahamstown in the 1850s to 1870s. Botanic gardens in Natal and the Cape Colony prominently featured Australian trees. Founded in 1881, the Cape Colony's Department of Forestry pursued the largest institutional programme of tree planting in southern Africa in the 1880s and 1890s. Natal, the Free State and the South African Republic all lagged behind the Cape in developing state departments of forestry. Except for Natal's failed attempt to maintain its fledgling Department of Forestry founded in 1891, none of these territories created a permanent department of forestry until after the South African War.[1] Private individuals, such as the British migrant into the Transvaal, Richard Wills Adlam (1853–1903), brought and planted the majority of Australian seeds in Natal, the Transvaal and the Free State.[2]

Australian botanists helped to direct many exchanges in the second half of the century. Australia's two most influential botanists in the late nineteenth century maintained extensive correspondence with scientists in southern Africa and helped settlers there select suitable species of trees to plant. Ferdinand von Mueller, the government botanist for Victoria from 1853 to 1896, sent seeds and provided advice to botanists, farmers and foresters for over 40 years (MacOwan, 1896: 627–628). Joseph Maiden, Director of the Sydney Botanic Gardens and Herbarium from 1896 to 1924, worked as the official seed collector for the Cape Colony from 1896, when the Agriculture Department, at the request of Hutchins, established a direct relationship with him.[3] Maiden continued in this role until 1905 when the Cape Government ended these exchanges to save money.[4]

Early diffusion: difficulty with Australian trees

The majority of Australian seeds entered southern Africa in the nineteenth century through the Cape Colony, though settlers in Natal and to a lesser extent the Free State and Transvaal all received and planted Australian trees (Beinart, 2003: 96). Cape settlers worked hand-in-hand with the government to plant Australian trees. One estimate suggested that the Department of Forestry sent out 300 million Acacia seeds alone to Cape colonists from 1882 to 1893 (Shaughnessey, 1980: 41; van Sittert, 2000: 660). Unfortunately, this massive distribution of seeds did not produce the desired results: problems with classification and provenance, a lack of prior experimentation, and unrealistic expectations led to disappointment in the last two decades of the nineteenth century when planted Australian seeds failed to grow (if they did at all) into the trees that the settlers had hoped they would.

To successfully grow an exotic tree first required that the planter knew the correct classification of the seed. But determining the botanical classification of Australian genera and species confounded settlers and even expert botanists. Eucalyptus classifications, in particular, proved troublesome. Genetic variations and local environmental influences can cause two trees of the same species to produce different leaf shape and growth forms. Australian collectors notoriously misclassified the species and provided poor geographical information on the regions from which they sourced seeds. Maiden discussed this problem candidly in a letter to the Agriculture Department of the Cape Colony: 'I cannot place your order in the hands of nurserymen, as their collectors are not at present sufficiently educated in regard to the difficult genus Eucalyptus to enable me to trust their naming.'[5] In another letter, Maiden told Hutchins to tell him if '[i]f any unusual proportion of the seeds fails to germinate, or if the seeds appear to be wrongly named, or to be under names different to those under which you have previously received them'.[6]

The same species planted in southern Africa often looked different than it did in Australia, making many published botanical guides inaccurate. 'No Genus is so perplexing as Eucalyptus in the matter of discrimination of species', the Cape Town botanist Peter MacOwan wrote, 'especially when as here, they have grown in fresh woods and pastures new, different from their Australian home, and have taken on a new habit' (MacOwan, 1893: 32). Questionable classifications led many settlers to call different species by the same name. MacOwan wrote an exasperated response to one settler who inquired about a specific species in 1894:

> The so-called popular names are the cause of endless wrangling and mis-understanding. Thus there are about twenty-five different White Gums, a dozen Blue Gums, several Black Wattles, several Golden Wattles, and every non-botanic grower vows that his particular blue or white or golden is the real one and the rest are bogus pretenders.
>
> *(MacOwan, 1894: 40)*

Settlers who planted Australian seeds did so with little knowledge of whether the species they (supposedly) selected would actually grow in the regions where they planted them. Trees could grow for a decade or more before showing signs of disease or other deficiencies. Hutchins noted that, '[most] trees, unless they are altogether unsuited to the climate and soil do well for a few years, perhaps the first 20 or 30 years' (Hutchins, 1905: 523). Inevitably, improperly classified seeds planted in climates vastly different from their native habitats rarely grew into healthy, valuable or useful trees. This led settlers to begin criticising eucalypts and some wattles.

Of all species planted widely in the mid to late nineteenth century, none attracted as much initial enthusiasm followed by criticism as *Eucalyptus globulus*. Starting in the late 1820s, settlers planted the species across all of southern Africa. The species was so widely planted in the nineteenth century that, despite not being planted for most of the twentieth century, many South Africans still call any species of Eucalyptus a 'blue gum' or 'bloekomboom'. The species grows best in a narrow climatic range similar to its cooler native habitats in New South Wales, Victoria, and Tasmania, but settlers planted it in deserts, on mountains, and in the subtropical interior. Even when trees did grow, the results could not live up to the high expectations stoked by enthusiasts. Farmers complained that their trunks twisted, rendering timber useless except for firewood (Ogston, 1903: 216). For a time, DeBeers did not buy them for use in the diamond mines (Farmer, 1903: 352). One farmer, P. H. Pringle, told readers of the *Agricultural Journal of the Cape of Good Hope* (*AJCGH*) how he had planted numerous genera and species of exotic trees, but found out that the 'least satisfactory of the lot is the Bluegum' (Pringle, 1903: 596).

The difficulties planters experienced with *Eucalyptus globulus* indicated a broader problem that scientists and settlers alike faced when selecting Australian genera and species of trees to plant. Rapid naturalisation under certain conditions of exotics such as *Acacia mearnsii* (black wattle) in the midlands of Natal in the 1880s and 1890s, happened largely as an accident, could not always easily be replicated on nearby sites and have historically overshadowed the widespread failure of trees planted in the nineteenth century (Macdonald *et al.*, 1986: 145–155). Another unsuccessful species was the *Eucalyptus robusta*, which settlers planted in the dry western districts of the Cape Colony. The seemingly healthy trees grew for a time, but '[t]hen came the inevitable failure. As a native of the damp semi-tropics of East Australia it was quite out of its place in the . . . climate of the Cape Peninsula or the dry Karoo' (Hutchins, 1905: 523). Although Eucalyptus species proved the most difficult to select, species of Australian Acacia, Araucaria, Casuarina and Hakea all provided problems for would-be planters in different parts of the colony. Would-be tree planters had little advice that detailed the native climates of an exotic species, let alone compared them with corresponding climates in the Cape Colony. Almost all tree-planting guides before the twentieth century lacked specificity, hindering the successful selection of species by settlers (see Storr Lister, 1884).

Creating exotic plantations at the Cape: climate and experimentation, 1881–1910

These failures encouraged foresters in the Cape Colony to begin thinking systematically about how to select exotic species. The creation of the Department of Agriculture and the Department of Forestry in the Cape Colony in the early 1880s provided government resources to pursue this research. With the publication of the Department of Forestry's first report in 1882, foresters began to keep track of experiments across the Cape. The *AJCGH*, the official mouthpiece of Department of Agriculture, gave regular updates on the results of experiments, offered a forum for debate, and helped to educate farmers in how properly to select, plant and maintain exotic trees. The Department of Forestry started experimental arboreta and plantations in the four forestry divisions of the colony.

Two foresters in particular guided the first three decades of research into exotics: Hutchins (Darrow, 1977) and Joseph Storr Lister. These two foresters played a much more influential role than did the Comte de Vasselot de Régné, the first conservator of the Cape Colony, who wrote and spoke in French and acted as little more than a 'figurehead' (Sim n.d.: 67). Storr Lister, unlike Hutchins, had no formal training in Europe. He was the first conservator born in southern Africa (in Cape Town in 1852). After working as a sub-assistant in the Punjab from 1871, Storr Lister returned to the Cape Colony to direct government drift-sand reclamation efforts at the Cape Flats. In 1883, the Department of Forestry, under the guidance of Storr Lister, founded the Tokai arboretum south of Cape Town. Lister promoted experimental planting throughout his career, which culminated in his appointment as the first Chief-Conservator of the Cape Colony in 1904 and the Chief-Conservator of the Union-wide Forestry Department in 1910. If Storr Lister acted as an institutional anchor that encouraged experiments with Australian trees, then Hutchins acted as the engine that drove forward actual experiments

Hutchins grew up in England and studied forestry at l'Ecole Nationale des Eaux et Forêts, in Nancy, France, in the early 1870s, before moving to southern India to work for the Indian Forest Service. He brought with him to the Cape an enthusiasm for Australian trees gained from working with them in 1881 around Ootacamund (Hutchins, 1883). Hutchins took the lead among Cape foresters in studying climate and experimenting with Australian trees. He adopted a straightforward concept to direct his efforts: 'fit the tree to the climate' (Hutchins, 1905: 521). When formulating his own climatic theories, Hutchins read widely, drawing heavily from meteorologists and botanists (especially those in Australia) rather than European foresters.[7] He argued passionately that knowing climatology and species' native climatic ranges were the most important subjects for foresters in the Cape to study. Rebutting James Currie, the Under-Secretary for Agriculture, who had denied his request for three books on Australian meteorology, Hutchins wrote, 'In South Africa with its variety of trees and climates, meteorology and the climate requirements of each tree are the most important study for foresters.'[8] Hutchins was an active Fellow of the Royal Meteorological Society, publishing an influential treatise that predicted climatic cycles based upon an analysis of sunspots and rainfall and temperature records (Hutchins, 1888: 37–112). He wrote articles and pamphlets for farmers recommending what species of trees to plant in specific regions of the Cape (Hutchins, 1902).

Most of Hutchins's ideas appeared in a multi-part essay, 'Extra-Tropical Forestry', published from 1905 to 1906 in successive issues of the *AJCGH*. Here, he argued that the Cape Colony had an 'extra-tropical' climate, meaning a climate predominating in regions near, but not in, the tropics, and characterised by dryness, abundant sunshine and variable rain. Cape colonists, Hutchins wrote, should select exotic trees from other extra-tropical regions. He noted specifically that there was a southern hemispheric extra-tropical zone, 'the sea-level climate between about latitude 23' and latitude 43' which embraced southern Africa, Australia, Argentina, southern Brazil, Chile and northern New Zealand' (Hutchins, 1905: 19). To pinpoint extra-tropical regions directly comparable to the Cape, he then analysed rainfall averages and patterns, altitude, average temperatures, light and humidity. He published no map of those regions from which his readers should select trees, but offered descriptive guidance. Hutchins was not alone in his prescriptions; the mantra 'fit the tree to the climate' dictated the Cape Colony's official forestry policy. C. B. McNaughton, a forester in the Cape Colony, similarly told farmers in 1904: 'Forest species may be grown far from their natural habitat provided that the local climate is similar to that to which they are naturally accustomed' (McNaughton, 1903: 4). The emphasis put Cape foresters at odds with many of their professional counterparts in northern Europe and Britain, who mocked Hutchins's seemingly eccentric interests in climate and experimental plantations.[9] Hutchins shrugged off these criticisms with the observation that:

> Forest Meteorology in Northern Europe is without the practical importance that it possesses in the extra-tropical parts of the world, and its study has been neglected in Europe, with the result that after the failure of many unsuitable trees, all introduced trees have been decried.
>
> *(Hutchins, 1905: 517)*

Meteorology and climatology mattered more in the Cape, he believed, because few indigenous forests existed and indigenous species could not be grown widely in plantations.

Hutchins's exuberant research into exotics struck the Director of Kew Gardens, William Thiselton-Dyer, as excessive and impractical. In 1896, Hutchins asked Kew to test some seeds from Europe before he purchased them. Thiselton-Dyer criticised his large, 10-ton order of seeds: 'I am obliged to remark that the instructions have been drawn up with want of practical knowledge ... some [species on the list] ... are actually unknown.'[10] Hutchins fired a letter back to the Secretary of Agriculture chiding Thiselton-Dyer: 'the remarks of the Director need not cause surprise. The Kew establishment can have had but a limited experience of the supply of forest seeds. Of forestry proper they have no knowledge either theoretical or practical.'[11] Hutchins confidently rebutted the most prominent botanist in the British Empire because he, like other Cape foresters, believed the Department of Forestry in the Cape Colony had a deeper knowledge of experimental plantings than foresters or botanists elsewhere in the world.

Hutchins may have had trouble with Kew, but he maintained a detailed correspondence with botanists in Australia, notably Maiden, who over a ten-year period offered him advice on what species to plant in the Cape Colony based upon their native climates. In their first exchange, Maiden decided to include *Eucalyptus saligna* (later properly identified as *Eucalyptus grandis*) in the shipment of seeds to Hutchins: 'It [*Eucalyptus saligna*] is not on your list, and you need not therefore pay for it unless you chose, but the expense is trifling.'[12] Hutchins asked to continue receiving this species, and Maiden's 'trifling' expense eventually became the most widely planted species of Eucalyptus in South Africa in the 1930s and after (Poynton, 1979b: 350–381).

South African research into climate and experimental plantations expanded from 1902 when Cape experts travelled north to start new departments of forestry: Thomas Sim went to Natal in 1902, K. A. Carlson to the Orange River Colony in 1903, and Charles Legat to the Transvaal in 1904. Hutchins officially toured the Transvaal after the war at the request of the reconstruction government to report on the extent of the territory's indigenous forests, to propose species of trees to plant, and to suggest areas to purchase land for plantations (Hutchins, 1903). He was unimpressed with what he saw: 'In tree-planting, the Transvaal, like others of the South African Colonies, has planted its trees entirely neglecting this most important consideration of climatic fitness' (ibid.: 124). He recommended planting eucalypts and Mexican pines in the Woodbush Range east of Pietersburg (Polokwane). Legat built on Hutchins's recommendations in the Transvaal, forming experimental plantations.[13] In the Orange River Colony, Storr Lister made recommendations on how to create a forestry programme and Carlson, the first conservator, actively promoted the creation of experimental plantations to find exotics that would grow there.[14]

From 1904 to 1906, conservators in each colony debated the need to create a forestry school in South Africa (Bennett, 2013). They agreed that the new school should emphasise local environmental conditions and pay attention to exotics and climate, subjects that they felt could not be properly learned in Britain, Europe or India. In 1906, a forestry school opened in conjunction with the South African College in Cape Town and the Cape Colony's Department of Forestry. Hutchins, who worked as the professor of forestry for the first year, emphasised climatology and exotics, an emphasis he hoped would lure students from Australia to attend the school. After Hutchins left for Kenya in 1907, foresters debated what type of professor should run the school (Darrow, 1977: 13). The Cape's Chief-Conservator, Storr Lister, worried about replacing Hutchins with a forester from India or Europe because he believed forestry in South Africa, rather than managing large forests, would focus on the 'formation and management of plantations of exotic trees and, notwithstanding past experience, Forest Officers for many years will have to continue to more or less feel their way by constant and systematic experiments'.[15] Thus, G. A. Wilmot, a South African who had studied forestry at Yale, directed the programme after Hutchins left.

Consolidation and growth, 1910–1948

In 1910, all four colonial departments merged into a single Union-wide Department of Forestry, which was centred in Pretoria but whose research agenda was dominated by Cape foresters for the next two decades (Dubow, 2006: 7–8). Storr Lister, the country's first Chief-Conservator, founded the department's Research Branch in 1912. When Storr Lister retired in 1913, Legat succeeded him as Chief-Conservator until 1931. Another former Cape forester, C. C. Robertson, joined the Research Branch in 1913 and started to direct experiments and climate research. Experiments became increasingly professional and centralised (Robertson, 1909: 219; Poynton, 1979b: 19). The Research Branch actively pursued studies in botany, silviculture, climatology, ecology, genetics and breeding, and the technological utilisation of timbers and bark.

The Cape lost its forestry school in Tokai when the South African College closed it down in 1911 after inter-colonial political tensions from 1906 to 1910 hindered the faculty's ability to select a suitable successor to Hutchins and enrolment faltered. The South African government decided instead to send forestry students to Oxford and elsewhere in Europe to study, and the country thus had no facility of its own to train officers until the opening of Stellenbosch University's department of forestry in 1932. Tokai's closure, however, had little effect on the ongoing research programme: except for a brief interlude from 1931 to 1935, every head of the Department of Forestry from 1910 to 1943 started their careers in the Cape Colony or had studied at the Tokai School.[16]

Working directly with the Research Branch, foresters in the field collected macro- and micro-climatic data countrywide to pinpoint sites and zones suitable for planting exotic tree species. The Annual Report of the Department of Forestry for 1931 published the first climatic silvicultural map of South Africa by dividing the country into zones according to temperature and rainfall.[17] This map provided a visual model that helped foresters understand the climatic zones of the country. At the same time, foresters kept detailed records of the topography, soil type, temperature, altitude, light and humidity of existing, classified experimental plantations. When combined, this macro- and micro-data allowed foresters to generate a more specific profile of which species to plant in any region and specific location in South Africa.

Foresters still struggled to classify Eucalyptus species growing in government plantations because of their unknown provenance, their similar characteristics and the tendency of certain eucalypts to hybridise.[18] Until the 1930s, few seeds sent from Australia came with detailed provenances (Poynton, 1979a: 15). After conservators suggested that the Department of Forestry send a forester to personally visit Australia, the Minister of Agriculture, Sir Thomas Smartt, chose Robertson, the researcher with the most knowledge of Australian trees in South Africa. In

1924, Robertson toured Australia's forests for six months studying its flora and climate. His report offered an analysis of similarities and differences between Australia and South Africa with recommendations on the habits and true classifications of species and genera (Robertson, 1926: 1). Robertson included a map that transposed the latitudes of each country onto each other as a way of demonstrating their similarities. The map did not explicitly recommend which regions to select species from (because, as Robertson noted, the countries' altitudes and climatic ranges varied), but it helped foresters to visualise more clearly how the two countries' climates potentially overlapped, and provided a rough guide of regions that might have climatically matched exotic species.

Ultimately, only experiments with a wide variety of species could truly show foresters the best species to plant. Most experiments led nowhere, but the few that succeeded laid the basis for plantation forestry in the twentieth century. Eventually, the results from experiments run from 1910 until the 1930s by the Department of Forestry helped inform foresters and private industry of exotic species to select and the methods used to plant, tend and harvest them. Some of the findings revealed the inadequacies of widely planted Eucalyptus species, especially *Eucalyptus globulus*, and opened the door for the use of other species, especially *Eucalyptus grandis* (then misidentified as *E. saligna*), *E. paniculata* and *E. maculata*.[19] Around the early 1930s, *Eucalyptus grandis* became the most widely planted Eucalyptus in South Africa. Experiments by I. J. Craib in the 1920s and 1930s helped to expand production of Acacia and pines (Hiley, 1959: 43).[20] The results of such experiments found direct application in the private plantation industry. Private plantations, especially of species of Eucalyptus in the Transvaal and Acacia and Eucalyptus in Natal, rapidly outpaced the size and profitability of their state counterparts (Reekie, 2004: 73–74).

From pinnacle to nadir, 1948–2010

Forestry research into climate and exotics reached its pinnacle between 1948 and 1994. During this period, the Nationalist government widely supported the expansion of private plantations of exotic trees. Forestry continued to evolve away from the Cape and into the Transvaal and Natal, the regions with the highest concentration of plantations of Australian trees. Government funds continued to support state research into selecting and testing exotic trees in sample plots and different sites. South African foresters continued to visit foreign countries and to comparatively study climates in order to find new species and better genetic strains of existing species to plant in South Africa (Loock, 1950). New theoretical modelling of climate suggested areas to plant exotics such as *Eucalyptus grandis* (Vowinckel, 1961: 91–104). By the early 1950s, following 70 years of research based upon 'a gradual process of elimination', foresters had become increasingly confident in their ability to list which species grew best where (King, 1951: 12–14; Poynton, 1959). Foresters recognised the lasting influence of Hutchins; a short biography was published on him as a Department of Forestry bulletin in 1977 (Darrow, 1977).

A large private plantation industry grew up as a result of increased knowledge about the ranges where Australian trees grew within South Africa. The earnings from products derived from plantations of Australian trees represented one of the fastest-growing parts of the South African economy from the 1960s until the early 1990s (Louw, 2004). The size of South Africa's Eucalyptus plantation estate continued to grow as a result of continued research and breeding programmes encouraged by an expanding market for eucalypt pulp that was dominated by the corporate conglomerates Mondi and Sappi. Plantations of eucalypts increased in size from 161,049 hectares in 1960 to 538,000 hectares in 1993.[21] Private Acacia plantations, primarily in Natal, reached their peak size in the 1950s at 350,000 hectares before slowly declining (Kull and Rangan, 2008: 1264). South African government foresters continued to plant wattle and eucalypts for rural development in the Bantustans, forcibly moving Africans to accommodate these plantations, fuelling their already strong resentment towards state forestry and exotic plantations (Tropp, 2003: 228).

The results from experimental arboreta and plantations across the country produced a large amount of data that had never been examined systematically. R. J. Poynton analysed much of this data when, drawing upon Department of Forestry records and published material, he issued his *magnum opus* of South African experimental silviculture in 1979, the massive two-volume *Tree Planting in South Africa* (Poynton,1968, 1979a; 1979b). Poynton intended his work as a guide to foresters on selecting exotic tree species, but it also very consciously reviewed the results of plantation experiments in South Africa and southern Africa (Angola, Malawi, Lesotho, Mozambique and Zimbabwe) that had been running since the late nineteenth century and were, he believed, coming to an end:

> Regarded from the historical viewpoint, this Report would seem to be opportune in as much that its publication comes more-or-less at the end of the era which witnessed the introduction of exotic forest trees in large numbers and their trial under varied conditions to determine their suitability or otherwise for timber production in Southern Africa. Although a few species of potential economic value doubtless still await testing, and while scope exists also for extending trials of established species to less favourable sites . . . henceforth the major advances in forestry research will be made in other directions.
>
> *(Poynton, 1979a: 815)*

Research, he predicted, would now focus on genetic provenance, breeding 'elite trees' and a more 'scientific approach towards matching the species to the site' (ibid.: 815). Hutchins's mantra, 'fit the tree to the climate', no longer served as the guiding light in forestry research in southern Africa.

A little over a decade later, the research programme that Poynton both epitomised and summarised derailed. The ending of apartheid opened up the floodgates of scientific and popular criticism of Australian trees. Not long after the

African National Congress took power in 1994, the environmental regulations regarding exotic plantations suddenly changed: the goal of water conservation, rather than that of economic growth, began to dictate the regulation of plantations. The newly created Department of Water Affairs and Forestry (DWAF) classified plantations of exotic trees as a 'stream flow reduction activity' that had to meet stringent criteria before being given government approval (Showers, 2010: 312). South Africa's plantation estate has stopped expanding and most growth is focused on less invasive, quick-growing eucalypts. As of 2007–2008, newly created plantations were composed primarily of Eucalyptus (74.3 per cent), with softwood pines (14 per cent) and Acacia (10.5 per cent) declining in importance (Republic of South Africa, Department of Agriculture, Forestry and Fisheries, 2008: xiii). Some 48.6 per cent of these eucalypts were E. grandis, the same species that Maiden sent to Hutchins in 1896. The amount of commercial plantation land under the direction of the government has plummeted to 62,000 hectares in 2010 and 100 per cent of all new plantations are created by the private sector (Department of Water Affairs and Forestry, 2009: 39).

A once powerful private and public forestry sector has taken a back seat to resurgent environmentalism and a new government emphasis on funding studies on native ecology and climate change. The public now has a 'neutral to negative' opinion of the forestry industry and forestry enrolments in universities have declined (ibid.: 41). In the place of forestry research, there has been a growing interest among environmental scientists in the study of 'invasive' and 'naturalised' Australian genera and species of trees. Australian trees are hunted down and exterminated by government employees funded by Working for Water, a programme founded in 1995 and funded by the DWAF (Neely, 2010: 869–887). Public opinion has swung in favour of these eradication programmes: popular discussions of Australian trees in newspapers sometimes display a powerful combination of ecological nationalism and xenophobia (Comaroff and Comaroff, 2001). Just as many nineteenth-century colonists enthusiastically wanted to believe only the best things about Australian trees, in a reversal of fortunes, many twenty-first-century South Africans want to believe only the worst.

Conclusion: the past and present

The currently popular anti-exotic rhetoric of many South Africans is at odds with the contribution of plantations and timber products to South Africa's economy and the more nuanced scientific findings about biological invasion held by the scientific community (van Wilgen, 2010). Australian trees retain a substantial ecological presence and remain an important part of the economy in South Africa. What happens to these trees is the topic of heated public and scientific debate. In 2011, the creation of plantations of exotic trees is seen as largely taboo by a South African public that seeks to protect its indigenous flora in order to promote tourism, preserve the nation's natural heritage and conserve critical water supplies. Yet Australian trees, even though many in the public strongly dislike them, will continue to be important economically, especially for residents in KwaZulu-Natal, Limpopo and Mpumalanga.

The very experiments that foresters used to successfully select species of Australian trees may hold the key to reversing the growth of these same trees or to identifying less invasive and thirsty species to grow in plantations. Ecologists who study the dynamics of biological invasion now draw upon the results of this climatic and experimental research to locate former arboreta and to assess the invasiveness of different species. There is still a vast amount of information on the history and growth potentials of individual species that can be gleaned from previous forestry research that has not been analysed. Future research from these sources has the potential to overturn common assumptions, reinforce current models and provide examples to develop new theories. We should remain open to what such a research programme might find. For now, what is clear is that the proliferation of Australian trees across South Africa did not happen by accident: foresters purposely studied, selected and planted exotics that successfully grew in South Africa. The results of their research are visible to anyone traversing the country today.

Notes

1 National Archives South Africa Pietermaritzburg [NASA-PMB], Colonial Secretary's Office [CSO], 1181, file 935; University of Cape Town Archives [UCT], Fourcade Bequest, BC 246, MSS C5, D.E. Hutchins to Henry Fourcade, 10 July 1890. In 1888, the Natal Government asked for and received information about the Cape's system of forestry. The Cape Colony sent the forester Henry Fourcade to Natal to report on their forests, but he decided not to stay and work as a forester there.
2 For a list of plants, see the papers of R. W. Adlam, UCT, R. W. Adlam, BC 815. Adlam moved to the Transvaal and served as the first curator of Joubert Park in Johannesburg from 1893 to 1903, helping to lay the foundation of a public garden in the city after the Transvaal's annexation in 1902. For Natal's wattle industry, see Witt (2005: 100–106). Ambiguity will always surround the introduction of many species into southern Africa because of the lack of accurate records and classifications.
3 See National Archives of South Africa Cape Town [NASA-CT], Department of Agriculture [AGR], 722, F719: Ernest Hutchins to Under-Secretary of Agriculture, 9 July 1896; Under-Secretary for Agriculture, Cape Colony, to Under-Secretary for Mines and Agriculture, New South Wales, 6 August 1897. Also see Frawley (2010), 'Joseph Maiden'.
4 NASA-CT, AGR 722, F719: Joseph Maiden to Under-Secretary of Agriculture, 3 November 1896; Under-Secretary for Agriculture to Agent General for the Cape of Good Hope, 7 June 1905.
5 NASA-CT, AGR 722, F719, Maiden to Under-Secretary of Agriculture, 3 November 1896.
6 NASA-CT, AGR 722, F719, Maiden to Under-Secretary of Agriculture, 13 August 1897.
7 For the books Hutchins ordered which discussed Australian meteorology and botany, see documents in NASACT, AGR 723, F791.
8 NASA-CT, AGR 723: B559/6, Hutchins to Under-Secretary of Agriculture, 13 December 1897; A27, Under-Secretary of Agriculture to Hutchins, 8 January 1898; B19/6/98, Hutchins to Under-Secretary of Agriculture, 20 January 1898 (quotation in latter). He also promoted this view publicly; see Hutchins (1905: 521).
9 See UCT, Fourcade Bequest, BC 246, C7, C. B. McNaughton to Henry Fourcade, 9 December 1909.

10 NASA-CT, AGR 725, No 117, Agent General of the Cape of Good Hope to Secretary for Agriculture, 14 September 1896, enclosing W. C. Thiselton-Dyer to Sir David Tennant, 12 September 1896.

11 NASA-CT, AGR 725, B566, Hutchins to Under-Secretary for Agriculture, 21 October 1896.

12 59 NASA-CT, AGR 722, F719, Maiden to Under-Secretary of Agriculture, 3 November 1896.

13 Transvaal Department of Agriculture, *Annual Report of the Director of Agriculture*, 1 July 1905 to 30 June 1906 (Pretoria, Government Stationery Office, 1907), pp. 14–15.

14 National Archives of South Africa Pretoria [NASA-P], Transvaal Agriculture Department [TAD], 540, 1181/06, H. F. Wilson to High Commissioner, 29 December 1904, enclosing J. Storr Lister, 'Forest Officers: Scientific Training Of', 29 November 1904; Carlson, *Transplanted: Being the Adventures of a Pioneer Forester in South Africa* (1947), Chapter 7.

15 NASA-P, TAD 540, 1181/06, F618, Joseph Storr Lister to Charles Legat, 6 January 1909.

16 J. D. Keet, Director of Forestry from 1935 to 1943 in the integrated Department of Agriculture and Forestry, studied at the South African College forestry school.

17 *Annual Report of the Department of Forestry for the Year Ended 31st March, 1931* (U.G. 11/1932), p. 28.

18 For some of the files related to classification and experiments of exotic species of tree, see NASA-CT: District Forest Officer Butterworth (FBT) 1/3; FBT 1/4; Chief Regional Forest Officer Transkei (FCT) 3/1/57; FCT 3/1/60; FCT 3/1/61. Foresters worked to create comprehensive guides of Eucalyptus species in South Africa. See Marsh, 'A Key to the Species of Eucalyptus Grown in South Africa' (1939), pp. 16–64.

19 See the research on *Eucalyptus saligna* from 1914 to 1923, documented in NASA-CT, FCT 3/1/61, T 953/51.

20 Interestingly, Hutchins made the suggestion to space pines wider apart, one of the major changes brought about by Craib's research. See NASA-CT, FCT 3/1/61, T953/74, C. E. Legat to all Conservators, 25 Feb. 1915. Craib's methods remained highly controversial in Europe until the 1960s.

21 For the 1960 statistics, see Republic of South Africa Department of Forestry, *Investigation of the Forest and Timber Industry of South Africa: Report on South Africa's Timber Resources, 1960* (1964), p. 7. For the 1993 statistic, see Davidson, 'Ecological Aspects of Eucalyptus Plantations' (1995/1996).

References

Barton, G. A. 2002. *Empire Forestry and the Origins of Environmentalism*, Cambridge, Cambridge University Press.

Beinart, W. 2003. *The Rise of Conservation in South Africa: Settlers, Livestock and the Environment 1770–1950*, Oxford, Oxford University Press.

Beinart, W. and Coates P. 1995. *Environment and History: The Taming of Nature in the USA and South Africa*, London, Routledge.

Beinart, W. and Middleton, K. 2004. Plant Transfers in Historical Perspective. *Environment and History*, 10, 3–29.

Bennett, B. M. 2010. The El Dorado of Forestry: The Eucalyptus in India, South Africa, and Thailand, 1850–2000. *International Review of Social History*, 55, 27–50.

Bennett, B. M. 2013. The Rise and Demise of South Africa's First School of Forestry. *Environment and History*, 19, 63–85.

Brown, K. 2001. The Conservation and Utilisation of the Natural World: Silviculture in the Cape Colony, c. 1902–1910. *Environment and History*, 4, 427–447.

Brown, K. 2003. 'Trees, Forests and Communities': Some Historiographical Approaches to Environmental History on Africa. *Area*, 35, 343–356.

Carlson, K. A. 1947. *Transplanted: Being the Adventures of a Pioneer Forester in South Africa*, Pretoria, Minerva Drukpers.

Carruthers, J. and Robin, L. 2010. Taxonomic Imperialism in the Battles for Acacia: Identity and Science in South Africa and Australia. *Transactions of the Royal Society of South Africa*, 65, 48–64.

Comaroff, J. and Comaroff, J. L. 2001. Naturing the Nation: Aliens, Apocalypse and the Postcolonial State. *Journal of Southern African Studies*, 27: 627–651.

Crosby, A. W. 1986. *Ecological Imperialism: The Biological Expansion of Europe, 900–1900*, Cambridge, Cambridge University Press.

Darrow, W. K. 1977. *David Ernest Hutchins: A Pioneer in South African Forestry*, Department of Forestry, Pretoria.

Davidson, J. 1995/1996. Ecological Aspects of Eucalyptus Plantations. In: *Proceedings: Regional Expert Consultation on Eucalyptus, 4–8 October 1993*, vol. 1, Bangkok, Food and Agricultural Organization. Available at: http://www.fao.org/docrep/005/ac777e/ac777 e06.htm#bm06 (accessed 30 July 2010).

Department of Water Affairs and Forestry. 2009. *Annual Report, 1 April 2008 to 31 March 2009* (R.P. 163/2009). Pretoria.

Drayton, R. 2000. *Nature's Government: Science, Imperial Britain, and the 'Improvement' of the World*, New Haven, CT, Yale University Press.

Dubow, S. 2006. *A Commonwealth of Knowledge: Science, Sensibility, and White South Africa, 1820–2000*, Oxford, Oxford University Press.

Farmer. 1903. De Beers and Blue Gum Wood, *Agricultural Journal of the Cape of Good Hope* 22: 352.

Frawley, J. 2010. Joseph Maiden and the National and Transnational Circulation of Wattle *Acacia* spp., *Historical Records of Australian Science*, 21: 35–54.

Grove, R. H. 1987. Early Themes in African Conservation: The Cape in the Nineteenth Century, in Conservation. In: Grove, R. and Anderson, D. (eds) *Africa: Peoples, Policies and Practice*, Cambridge, Cambridge University Press, pp. 21–40.

Grove, R. H. 1995. *Green Imperialism: Colonial Expansion, Tropical Island Edens and the Origins of Environmentalism, 1600–1860*, Cambridge, Cambridge University Press.

Hiley, W. E. 1959. *Conifers: South African Methods of Cultivation*, London, Faber and Faber.

Hooker, J. D. 1853. *Flora Novae Zelandiae*, London, Lovell Reeve.

Hooker, J. D. 1860. *Flora Tasmaniae*, London, Lovell Reeve.

Hutchins, D. E. 1883. *Report on Measurements of the Growth of Australian Trees on the Nilgiris*, Madras, Government Press.

Hutchins, D. E. 1888. Cycles of Drought and Good Seasons in South Africa, *Wynberg Times*, Wynberg.

Hutchins, D. E. 1902. *A Chat on Tree Planting with Farmers*, Cape Town, W. A. Richards and Sons.

Hutchins, D. E. 1903. *Transvaal Forest Report*, Pretoria, Government Printing and Stationery Office.

Hutchins, D. E. 1905. Extra-Tropical Forestry: Being Notes on Timber and Other Trees Cultivated in South Africa and in the Extra-Tropical Forests of Other Countries. *Agricultural Journal of the Cape of Good Hope*, 26: 18–19.

King, N. L. 1951. Tree-Planting in South Africa. *Journal of the South African Forestry Association*, 21, i–102.

Kull, C. A. and Rangan, H. 2008. Acacia Exchanges: Wattles, Thorn Trees, and the Study of Plant Movements. *Geoforum*, 39, 1, 258–272.

Loock, E. E. M. 1950. *The Pines of Mexico and British Honduras: A Report of a Reconnaissance of Mexico and British Honduras during 1947*, Pretoria, Government Printer.

Louw, W. J. A. 2004. General History of the South African Forest Industry: 1975 to 1990. *Southern African Forestry Journal*, 200, 77–86.

Macdonald, I. A. W., Kruger F. J. and Ferrar, A. A. 1986. Processes of Invasion by Alien Plants. In: Macdonald, I. A. W., Kruger F. J. and Ferrar, A. A. (eds) *The Ecology and Management of Biological Invasions in Southern Africa*, Cape Town, Oxford University Press, pp. 145–155.

MacOwan, P. 1893. Gum of Eucalyptus. *Agricultural Journal of the Cape of Good Hope*, 6, 32.

MacOwan, P. 1894. Australian Hedge Plant. *Agricultural Journal of the Cape of Good Hope*, 7, 40.

MacOwan, P. 1896. The Late Baron Sir Ferdinand von Mueller, C.M.G., F.R.S., Etc., Government Botanist of Victoria. *Agricultural Journal of the Cape of Good Hope*, 9, 627–628.

Marsh, E. K. 1939. A Key to the Species of Eucalyptus Grown in South Africa. *Journal of the Southern African Forestry Association*, 3, 16–64.

McCracken, D. 1997. *Gardens of Empire: Botanical Institutions of the Victorian British Empire*, London, Leicester University Press.

McNaughton, C. B. 1903. *Tree Planting for Timber and Fuel*, Cape Town, Townshend Taylor and Shashall.

Mucina, L. and Rutherford, M. C. (eds) 2006. *The Vegetation of South Africa, Lesotho and Swaziland*, Pretoria, South African National Biodiversity Institute.

Neely, A. 2010. 'Blame it on the Weeds': Politics, Poverty, and Ecology in the New South Africa. *Journal of Southern African Studies*, 36, 869–887.

Noble, J. 1886. *History, Productions, and Resources of the Cape of Good Hope*, Cape Town, W. A. Richards and Sons.

Ogston, E. E. 1903. The Twisting of Blue Gums. *Agricultural Journal of the Cape of Good Hope*, 22, 216.

Pooley, S. 2009. Jan van Riebeeck as Pioneering Explorer and Conservator of Natural Resources at the Cape of Good Hope (1652–62). *Environment and History*, 15, 3–33.

Pooley, S. 2010. Pressed Flowers: Notions of Indigenous and Alien Vegetation in South Africa's Western Cape, c. 1902–1945. *Journal of Southern African Studies*, 36, 599–618.

Poynton, R. J. 1959. *Notes on Exotic Forest Trees in South Africa*, 2nd edn, Pretoria, Government Printer.

Poynton, R. J. 1968. Trees for the Western Transvaal Selected on the Basis of Arboretum Trials, MA thesis, Stellenbosch University.

Poynton, R. J. 1979a. *Tree Planting in Southern Africa*, vol. 1, *The Pines*, Pretoria, Department of Forestry.

Poynton, R. J. 1979b. *Tree Planting in Southern Africa*, vol. 2, *The Eucalypts*, Pretoria, Department of Forestry.

Pringle, R. H. 1903. Blue Gums as Fuel and Timber Trees. *Agricultural Journal of the Cape of Good Hope*, 22, 596.

Radkau, J. 2008. *Nature and Power: A Global History of the Environment*, New York, Cambridge University Press.

Rajan, R. 2006. *Modernizing Nature: Forestry and Imperial Eco-Development 1800–1950*, Oxford, Oxford University Press.

Reekie, W. D. 2004. The Wood from the Trees: Ex Libri ad Historiam Pertinentes Cognoscere. *South African Journal of Economic History*, 19, 73–74.

Republic of South Africa, Department of Agriculture, Forestry and Fisheries. 2008. *Report on Commercial Timber Resources and Primary Roundwood Processing in South Africa, 2007/ 2008*. Available at: http://www.forestry.co.za/uploads/File/industry_info/statistical_ data/Timber%20 Statistics%20Report%202007_2008%20Final.doc.

Republic of South Africa Department of Forestry. 1964. *Investigation of the Forest and Timber Industry of South Africa: Report on South Africa's Timber Resources, 1960*, Pretoria, Government Printer.

Robertson, C. C. 1909. Some Suggestions as to the Principles of the Scientific Naturalisation of Exotic Forest Trees. *South African Journal of Science*, 6, 219–230.

Robertson, C. C. 1926. The Trees of Extra-Tropical Australia: A Reconnaissance of the Forest Trees of Australia from the Point of View of Their Cultivation in South Africa, *Cape Times Limited*, Cape Town.

Shaughnessey, G. L. 1980. Historical Ecology of Alien Woody Plants in the Vicinity of Cape Town, South Africa, PhD thesis, University of Cape Town.

Shaughnessey, G. L. 1987. A Case Study of Some Wood Plant Introductions to the Cape Town Area. In: Macdonald, I. A. W., Kruger, F. J. and Ferrar, A. A. (eds) *The Ecology and Management of Biological Invasions in Southern Africa*, Oxford, Oxford University Press, pp. 37–43.

Showers, K. B. 2010. Prehistory of Southern African Forestry: From Vegetable Garden to Tree Plantation. *Environment and History*, 16, 295–322.

Sim, T. R. Autobiography, unpublished report, Stellenbosch University Engineering and Forestry Library, MSS.

van Sittert, L. 2000. 'The Seed Blows About in Every Breeze': Noxious Weed Eradication in the Cape Colony, 1860–1909. *Journal of Southern African Studies*, 26, 655–674.

van Sittert, L. 2004. The Nature of Power: Cape Environmental History, the History of Ideas and Neoliberal Historiography. *The Journal of African History*, 45, 305–313.

Storr Lister, J. 1884. *Practical Hints on Tree Planting in the Cape Colony*, Cape Town, W. A. Richards and Sons.

Tomlinson, B. R. 1988. Empire of the Dandelion: Ecological Imperialism and Economic Expansion 1860–1914. *Journal of Imperial and Commonwealth History*, 26, 84–99.

Tropp, J. A. 2003. Displaced People, Replaced Narratives: Forest Conflicts and Historical Perspectives in the Tsolo District, Transkei. *Journal of Southern African Studies*, 29, 207–233.

Tropp, J. A. 2006. *Natures of Colonial Change: Environmental Relations in the Making of the Transkei*, Athens, OH, Ohio University Press.

Vandergeest, P. and Peluso, N. L. 2006. Empires of Forestry: Professional Forestry and State Power in Southeast Asia, Part 1. *Environment and History*, 12: 31–64.

Vowinckel, E. 1961. Potential Growth Areas for Introduced Tree Species. *Forestry in South Africa*, 1: 91–104.

Wallace, A. R. 1880. *Island Life*, London, Macmillan.

van Wilgen, B. 2010. Be Afraid, the Invasion is Real. *Witness*, 10 November.

Witt, H. 2002. The Emergence of Privately Grown Industrial Tree Plantations. In: Dovers, S., Edgecombe, R. and Guest, B. (eds), *South Africa's Environmental History: Cases and Comparisons*, Athens, OH, Ohio University Press, pp. 90–111.

Witt, H. 2005. 'Clothing the Once Bare Brown Hills of Natal': The Origin and Development of Wattle Growing in Natal, 1860–1960. *South African Historical Journal*, 53: 99–122.

14

REMAKING WETLANDS

Rice fields and ducks in the Murrumbidgee River region, NSW

Emily O'Gorman

In 1932, James Roy Kinghorn, a zoologist at the Australian Museum in Sydney, published the results of his investigation into whether ducks damaged rice crops in the Murrumbidgee Irrigation Area (MIA), a farming settlement that drew water from the Murrumbidgee River in inland NSW (Figure 14.1). He began by explaining the background to this research, which focused on two rice seasons, in the summers of 1926–1927 and 1927–1928. During these years, widespread drought conditions had attracted 'thousands of wild ducks and other waterfowl' (1932: 603) to the region. In the first of these seasons, some growers blamed their reduced crop on duck damage, arguing 'that both the black duck and the grey teal ate the freshly sown seed and sprouting plants, puddled the bays (thus interfering with the young plants), and later destroyed the rice when it was in head'.[1] The drought continued into the following rice season and the ducks stayed. Fearing more damage, a group of farmers attempted 'to have the names of several species of ducks removed from the list of protected birds' during an enforced closed season so that they could be hunted as pests. However, other farmers claimed that the ducks had done almost no damage and in fact helped their farming by eating weeds in the rice bays. Because of this 'contradictory evidence, and as there was no information available as to the economic value of wild ducks in regard to rice cultivation', Kinghorn was 'asked to investigate'. He initially tried to undertake this research from Sydney, by examining the gizzards of 17 'ducks forwarded from the Area', of four different species. However, he became wary of this proof, 'as it was evident that some of the ducks, prior to being killed had been forcibly or purposefully fed with mature rice grains' (1932: 603. See also Ellis, 1940: 201–202; Correspondence, 1926, Australian Museum) (Figure 14.2). Kinghorn therefore decided to make his own field investigations. This was only the third season of a commercial rice crop in the MIA and controversies around ducks in rice fields had already begun.

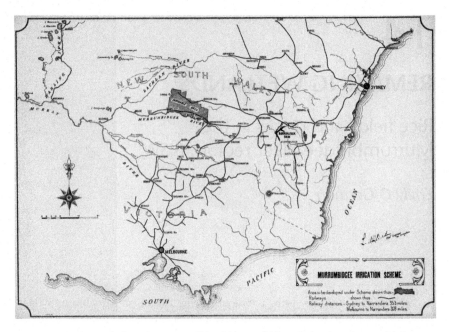

FIGURE 14.1 An early location map of the Murrumbidgee Irrigation Area, which officially began operation in 1912.

Source: State Records NSW: NRS 14086, Lantern slides of NSW and the Franco-British Exhibition, 1905–11. Digital ID: 14086_a005_a005SZ847000020, Map of southern NSW showing Murrumbidgee Irrigation Area, 1908.

Native species of ducks continued to visit the rice fields and they have remained controversial figures in the region. They have occupied a complex and multifaceted role for managers, biologists, and farmers: variously portrayed as wild or native and so belonging, or as agriculturally disruptive and invasive, they have been blamed for damaging rice crops by some farmers (but not all) and hunted as pests, while others have valued them as game birds and for their role in eating invertebrates that damage crops. Ducks have therefore brought together many different sets of interests and long historical legacies such as maintaining populations for hunting, the conservation of wildlife, and agricultural economics, all of which came to bear in the controversies over whether or not they damaged rice crops and how they should be treated in these areas.

This chapter examines how these competing ideas built up around ducks in the first half of the twentieth century as commercial rice growing was established and expanded in this region, which soon became the centre of Australia's rice industry. I particularly focus on the involvement of two biologists, Kinghorn in the 1920s and 1930s and Harry Frith in the 1950s, who bring together these different interests in ducks. Both undertook government research, generated by farmers' concerns, into whether ducks damaged rice crops. This research reflected a dominant focus throughout the twentieth century in Australian government science on agricultural

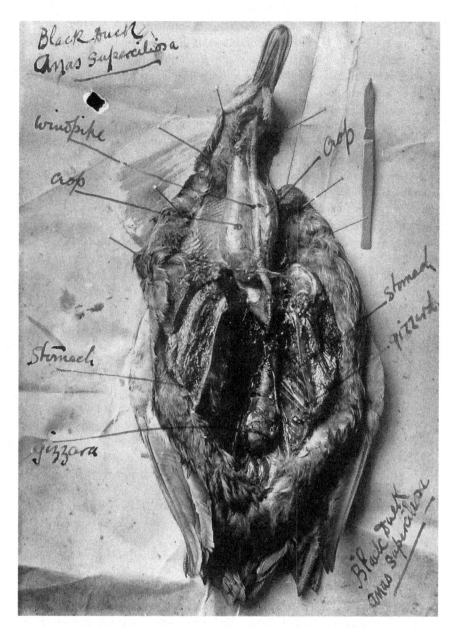

Black Duck.
Anas superciliosa

windpipe

crop

crop

stomach
gizzard

Stomach

gizzard

Black duck
anas superciliosa

FIGURE 14.2 A dissected Pacific 'Black Duck'. This image was included in J. R. Kinghorn's correspondence at the Australian Museum, in which he made some of the practical arrangements for obtaining dead ducks from the Murrumbidgee Irrigation Area, in order to discover whether or not they had eaten rice crops. This image was likely used to demonstrate duck anatomy and may not be one of the ducks sent from the Murrumbidgee Irrigation Area to Kinghorn, though it is one of the same species.

Source: J. R. Kinghorn correspondence. AMS9/ C80/1926. Courtesy Australian Museum Archives.

economics, a key part of which was research on pests (Robin, 1997: 69–73; Dunlap, 1997: 77–81). The work and lives of these two biologists also provide lenses into changing ideas of pests and conservation: from Kinghorn's zoological research with the Australian Museum to Frith's explicitly ecological work for the Wildlife Survey Section of the Commonwealth Scientific and Industrial Research Organisation (CSIRO). Through their research, we can gain insight into some of the complexities at the interface between wildlife and agriculture as each negotiated conflicting understandings of how ducks could and could not be accommodated within farmland.

Most importantly, this chapter examines some of the changes brought on by the intensification and expansion of irrigated agriculture in the twentieth century, which by degrees changed the water landscapes of the Murrumbidgee, Australia and, indeed, the world. Since rice was first grown commercially under extensive irrigation in the MIA, water diversion for flooded rice paddy irrigation has meant that these agricultural landscapes have sometimes 'replaced' original wetland habitat for animals such as ducks. The expansion of agriculture changed wetland environments and so reconfigured possibilities for animals and plants, with a range of mixed outcomes for all those involved.

Background: Murrumbidgee River, ducks, and rice

The Murrumbidgee River is a major tributary of the Murray River, located in the south-east of the Murray–Darling Basin. Like other rivers in the Murray system, snowmelt in the southern parts of the Great Dividing Range contributes to seasonal inflows in the winter and spring months and sometimes to large floods when these are accompanied by substantial rain. The Murrumbidgee and its tributaries can, however, flood at any time from heavy rainfall, as they did in March 2012. During floods, biodiversity and populations of particular species increase in wetlands, along rivers and on floodplains, as floods trigger breeding cues for many plants and animals. The Murrumbidgee River and its floodplains can also experience periods of intense drying, lasting years, as they did in some places in the region from 2000 to 2010 (Pittock *et al.*, 2006: 17–18; Drendel, 2011).

The populations and movements of most species of native ducks are considerably influenced by these wet and dry periods. These ducks are, in varying degrees, nomadic; that is, rather than having a seasonal migration, for example, as ducks in North America do, they flock to rivers and wetlands that are in flood, to feed, nest and breed. The species of duck that visit the Murrumbidgee region include the Grey Teal (*Anas gracilis*), which are extremely nomadic and which can fly large distances to flooded waterways and wetlands (Roshier, 2009: 76–78; Kinghorn, 1932: 603; Frith, 1957a: 33; Curtin and Kingsford, 1997: 3). Other species are the Pacific Black Duck (*Anas superciliosa*); the Pink-eared Duck (*Malacorhynchus membranaceus*); the Australian Wood Duck (Maned Goose, *Chenonetta jubata*, which is actually closer to species of geese than ducks); the Australian Shelduck (Mountain Duck, *Tadorna tadornoides*); and the Plumed

Whistling-duck (Plumed-tree Duck, *Dendrocygna eytoni*), among others. The populations of each of these species significantly increase during large floods as females hatch several broods in a few months and crash during droughts as breeding slows. While breeding can occur at any time of year, spring river rises in the Murrumbidgee and Murray mean ducks often breed there in these months (Curtin and Kingsford, 1997: 9; Kingsford and Norman, 2002; Roshier *et al.*, 2002). Local Aboriginal peoples along the Murrumbidgee, Wiradjuri and Ngunawal, have traditionally hunted these ducks and eaten their eggs, particularly during floods (Humphries, 2007: 107).

Many large dams and other works were built in the Murray system in the twentieth century in attempts to control the variable flows of the rivers, largely for agriculture (Lloyd, 1988: 181–184; Connell, 2007: 56–61). Today, the Murray–Darling Basin produces a significant amount of Australia's food for export and domestic consumption (Australian Bureau of Statistics, 2008). The construction of large dams and the spread of irrigation significantly altered many of the rivers' flows and in recent years water and land management in this area has become an increasingly controversial topic (Kingsford, 2000). In broad terms, the most prevalent issue is perhaps the conflict between the water needs of established agricultural farms and industries, and the ramifying ecological degradation from these water extractions, with consequences for downstream farmers and Indigenous cultures and livelihoods (Weir, 2009: 26–46; Sinclair, 2001: 3–25).

The Murrumbidgee River has been central in these recent debates. The river supports extensive irrigation networks and is also regulated by a number of large dams, including Burrinjuck Dam and those in the Snowy Mountains Hydro-Electric Scheme. In recent years almost all rice grown in Australia has been grown in the three irrigation areas that are located in the Murrumbridgee and Murray region: the MIA, the Coleambally Irrigation Area, and the Murray valley irrigation districts (Australian Bureau of Statistics, 2005). The main varieties of rice grown in these areas are medium-grain Japonica, which are traditionally flood-irrigated (in Australia and elsewhere). The semi-aquatic characteristics of rice mean that it needs to be flooded at particular times in its growth cycle to ensure high yields and to control weeds. The relatively high water requirements of the crop have led to its cultivation being heavily criticised within broader national debates ('Environment Committee' and 'Press Cuttings 1', Ricegrowers' Association of Australia; Barclay, 2010). In addition, there are a number of wetlands along the Murrumbidgee, such as the Lower Murrumbidgee floodplain, that are important sites for bird breeding and biological diversity more broadly, which have received reduced amounts of water due to upstream irrigation extractions (Kingsford and Thomas, 2004: 286–294). The loss of wetland habitat in the last three decades has led to an overall decline in numbers of some duck species like Grey Teal throughout eastern Australia. Yet, the changing water landscapes of the Murrumbidgee and other river regions have also been inhabited by water-birds in a range of different ways. For instance, populations of Wood Duck and Pacific Black Duck have increased in recent years, partly through their inhabitation of increasing numbers of farm dams (Curtin and Kingsford, 1997: 32; Kingsford *et al.*, 2000: 4).

Similarly, as we have seen, ducks have been attracted to flooded rice fields both as drought refuges and as aquatic habitat during local floods. In the words of ecologists Alison Curtin and Richard Kingsford, '[w]hen rice is flooded, it creates a wetland' (Curtin and Kingsford, 1997: 31). However, the sharing of space and resources by agriculture and wildlife in Australia has rarely been a matter of easy coexistence as many species, like ducks, have been and continue to be hunted or targeted in other ways as agricultural pests (see Main, 2005: 16–56). In the late 1990s, the irregular and large variations in duck populations, which change with droughts and floods, were put forward by managers as a major reason why management solutions for sharing these landscapes could not been found (Curtin and Kingsford, 1997: 9–10). This was in contrast to the more effective management in other countries (albeit contested and highly politicised), like the creation of wetland reserves in Californian rice growing areas, which catered for the more regular seasonal migrations of waterfowl (Wilson, 2012: 10–15).

All of these recent circumstances are underlain by complex histories, and are tied to the intensification and expansion of agriculture in the twentieth century. While rivers had been dammed and diverted before this, the rapidity and scale of agricultural expansion in the twentieth century were new and in that century many of the large dams that now stand in the Murray–Darling Basin were built; in the year 2000 these numbered 105 (Kingsford, 2000: 118). The Murrumbidgee was an important site for these projects and, as historical geographers Trevor Langford-Smith and John Rutherford noted, by the mid-1960s the Riverine Plain had 'by far the greatest concentration of irrigation development' (1966: 1) in Australia. This expansion was motivated by a range of political and cultural ideas and events. At the turn of the twentieth century these included: long-held ideologies of 'closer settlement' (or densely settled farming communities) for more intensive production, particularly of food; the economic fallout of what is now known as the Federation Drought (approximately 1895 to 1902); and the new legal framework of the national Constitution following Federation in 1901, which facilitated co-operation between different states for river engineering along the shared waterways of the Murray system (for a more detailed overview, see O'Gorman, 2012: 119–134).

Rice and the Murrumbidgee Irrigation Area

Within these contexts, the MIA began operation in 1912 as a state government-administered irrigation area that was intended to produce food from small, inten-sively cultivated farms, and the original production plans included horticulture, dairying, lucerne and raising pigs and fat lambs. The scheme was motivated by broader ideologies held by Australian governments of creating a 'yeoman' class of small-scale farmer and these racially-motivated ideas about 'desirable' immi-grants were reflected in the recruitment of farmers from the USA and Europe, as well as elsewhere in Australia (Langford-Smith and Rutherford, 1966: 25–29,

33–43; Powell, 1989: 41; Lloyd, 1988: 199–212; Waterhouse, 2005: 60–66, 200–201).[2]

The establishment of the MIA created further upheaval for Wiradjuri. Following the often violent frontier conflicts between mostly Anglo-Celtic colonists and Wiradjuri, which reached their height in the late 1830s, many Wiradjuri had taken up employment on the pastoral properties that dominated the region from the 1840s, mostly as a means to stay on country. In the first two decades following the establishment of the MIA, more Wiradjuri moved into Missions or to places outside these intensively cultivated regions (Gammage, 1983: 3–17; Kabaila, 1995: 117–141).

As farmers began cultivation, it soon became clear that many of the crops were unsuitable for intensive farming in the MIA. Soil salinity became a particular problem, and was so severe in some places that farmers abandoned their blocks (Langford-Smith and Rutherford, 1966: 25–29, 33–43; Lloyd, 1988: 199–212). With the MIA bordering on failure, rice became one of a number of agricultural products under trial by farmers and the state government to replace the unsuccessful products. Initial trials of rice were undertaken at the nearby Yanco Experiment Farm in 1915 by the NSW Department of Agriculture. These varieties had been bred from Japanese rice by a recently migrated couple, Isaburo (Jō) and Ichiko Takasuka. These trials were promising but not on the whole successful, and it was not until Californian varieties were trialled between 1920 and 1923 that rice moved into commercial production, with the first commercial crop being planted in 1924 (Langford-Smith and Rutherford, 1966:31; Watkins, 1926: 748; NSW Department of Agriculture, c.1948: 2; 'Agriculture', 384A, SRNSW). Despite its water usage, rice was seen to be valuable by farmers and the state government because in many ways it suited the clay soils that underlay parts of the MIA that had previously been seen as unproductive. The clay soils held water well and were dense enough to prevent water tables from rising and thereby causing salinisation. Rice cultivation also created suitable conditions for mixed farming, as the water that was still held in the soil following a rice crop could be used to grow other crops or fodder for livestock (Watkins, 1926: 748; 'Integration', SRNSW).

Ducks and the establishment of rice: Kinghorn

During the first commercial season, farmers reported damage to rice crops to the Australian Museum, requesting a special open season on ducks (McKeown, 1923: 43; Frith, 1957a: 33; Correspondence, Australian Museum). The Australian Museum was established in the early nineteenth century as a natural history museum and its primary purpose was the collection of animal specimens and taxonomic research. In the early twentieth century parts of the Museum became redirected towards the needs of agricultural departments, as at other research institutions in Australia (Kohlstedt, 1983; Robin, 1997). For instance, the Museum undertook applied research into animal biology and behaviour that related to agriculture, which drew on its strengths in zoology.[3] The Museum recommended to the Chief Secretary – who declared open seasons – that the open season currently in place in the region, from February to April, be maintained and not altered to better suit the times when rice was vulnerable, from approximately September to March (Ellis, 1940: 201). This was possibly due to concerns that hunting across the summer months would coincide with the peak of the duck nesting season. Duck breeding seasons had traditionally been closed hunting seasons, so as to protect duck populations for game shooting. Another reason may have been due to some farmers' assertions that the birds ate insects that damaged rice crops (Kinghorn, 1932: 603).

Dry conditions in the mid-1920s led the birds to seek refuge in the rice fields. Rising controversy over whether they were a pest that hindered the rice crop or helped by eating weeds sparked the government investigations by Kinghorn. Kinghorn visited the MIA three times during the 1927–1928 rice season, at different stages of cultivation and plant growth. During this time he talked to farmers, observed birds on rice fields, and hunted ducks in rice areas and nearby wetlands such as Fivebough Swamp for his own gizzard analyses (totalling 17 ducks from four different species). Kinghorn's investigations reveal his unfolding views about ducks and agriculture. They also show some of the different views and practices of farmers and the ways that ducks and other birds were inhabiting these new aquatic environments (1932: 604–606).

Kinghorn found that grey teal, black duck, and wood duck ate rice seed at the time of sowing as well as green rice plants, but also barnyard grass-weeds, 'weed seed' and some insects. Pink-eared duck, however, ate no rice and were instead valuable 'as an insect destroyer' (ibid.: 607). He also argued that other waterbirds like the glossy ibis, cranes, herons, white ibis, and spoonbills were helpful to rice farmers as they ate insects and crayfish, the latter of which could undermine check banks, drains and other earthen infrastructure. Further, Kinghorn found that ducks were not the only birds that potentially damaged rice crops. As the first flooding took place in the rice bays, he saw starlings and crows 'retreating before the advancing water collecting worms and insects which were wriggling hurriedly from the soil', 'but at the same time there appeared to be no doubt that they were also eating the freshly sown rice seed' (ibid.: 604). It seemed that ducks had been blamed for the damage these, and other birds like galahs and sparrows, had caused as ducks were considered by some farmers to be 'the only bird' to damage rice (ibid.: 606).

While ducks puddled and pulled out some rice, Kinghorn argued that much of the damage attributed to them was instead due to 'faulty farming methods' (ibid.: 608). He reached this conclusion through his interviews with farmers where he found that '[i]t was also very noticeable that the growers who produced the best rice crops had little or nothing to say against the ducks' (ibid.: 607). Kinghorn claimed that bare sections in a rice crop were mostly due to the rice, which had been sown by being broadcast, washing into depressions and then drowning as water pooled there when the rest of the bay was drained. Similarly, rice could drown if the bays were badly graded. Broadcasting seed also seemed to produce a weaker young rice plant than drill sowing. Plants grown from broadcast seed could therefore be more easily puddled by ducks and were often washed onto the banks by wind (ibid.: 608).

The main conclusion of Kinghorn's report was that overall 'wild ducks' were 'not a serious pest of rice crops'. While ducks as well as other '[n]ative companions' could do some damage if they descended in large numbers during droughts, he found that it was 'impossible to say' the extent of this damage, as there were many other factors, and critters, involved (ibid.: 607–608). Kinghorn pointed to a complex situation emerging in rice areas, which was not just about ducks and rice but many other connections and relationships.

While we can perhaps recognise an ecological approach in Kinghorn's study, he did not couch his research in theses terms. Ecology was not then being widely taken up as an applied science and there is no evidence that Kinghorn took a particular interest in the emerging field (Dunlap, 1997: 76–78; Robin, 1997: 68–70;). Kinghorn's approach is perhaps better understood through changing approaches in field biology and his research echoes other field studies being undertaken around the world at the time. For instance, Herbert Stoddard's work on the decline of quail numbers in Georgia from 1924 to 1929, undertaken for the US Agricultural Department's Bureau of Biological Survey at the request of hunters, took a broadly comparable approach (albeit to a very different question), and Stoddard also claimed that a central factor was farming practices, which adversely affected the quail (Dunlap, 1997: 77–78). Historian Thomas Dunlap has argued that Stoddard's research was not explicitly ecology but that it nevertheless 'was marked by a new view of nature . . . [where] Nature was not observed, it was constructed' (1997: 77–78). Similarly, Kinghorn placed people within the landscape in a relatively new way by arguing that a key issue was farming practices.

Kinghorn thought that the belief by some farmers that ducks were responsible for significant rice damage was mostly 'the outcome of imagination and founded on hearsay' (1932: 606). Perhaps preempting a backlash by farmers, in his unpublished report (which appears to have been available to growers), Kinghorn wrote:

> for the information of the growers, I would like to say that I went into this matter absolutely unbiased, and in the early days of the investigation was inclined to believe ducks were doing a lot of damage. As time advanced my opinion changed and I was eventually able to prove to my satisfaction that the alleged damage was exaggerated.
>
> (unpublished report quoted in Ellis, 1940: 203)

Yet, enough farmers maintained that ducks did do appreciable damage that after further complaints in 1932, the NSW government allowed farmers to apply to shoot ducks, waterhens, and red-bills, on their property from August to December. However, Frith later indicated that 'only six applications for permits were received' (1957a: 32).[4]

Kinghorn's findings may have been dismissed as 'top-down' science, which was often resented by farmers. Further, farmers may have disagreed with Kinghorn about what constituted 'serious' or acceptable damage. For some farmers, having any losses to ducks or having any animal seen to be a pest on their farm could have been seen as unacceptable. In some cases, this approach seems to have been underpinned by a dominating view of the landscape in which farmers had ultimate control over what did and did not 'fit' on their farms (see Rose, 2004: 53–72; Rose, 2006: 66–68; van Dooren, 2011: 290). That is, if ducks were not seen to belong on rice farms, then they would be forced off the farms or killed.

Native pests, the economic value of birds, and hunting

These debates about ducks in rice fields draw attention to some of the wider histories of how people have understood the relationships between native animals and agriculture, particularly competing economic views of some native animals as pests or as helpful to farming through their eating of weeds and other animal and insect pests. In NSW, the destruction of native and introduced fauna that were regarded as agricultural pests was legislated for in 1880 with the *Pastures and Stock Protection Act*. This Act initiated a system of bounties on particular native and introduced animals that lasted for 50 years, but did not include ducks or other birds (Stubbs, 2001: 29–30; Jarman and Brock, 2004: 4). From the mid-nineteenth century through to the early twentieth century, birds were largely included in legislation under 'game laws' that intended to conserve game populations by limiting over-hunting, initially through a series of years of complete bans on hunting and then through the introduction of 'open seasons' that aimed to protect birds during their breeding seasons. Many native ducks, including those that later visited rice fields, took on the role of similar species in European hunting activities and were included under these laws as game birds (Walker, 1991: 18; Bonyhady, 2000: 14–39; Stubbs, 2001: 26–28; Jarman and Brock, 2004: 2; Dow, 2008: 148). The first of these Acts, the *Game Protection Act 1866*, explicitly exempted two groups of people: government approved collectors of natural history specimens and Aboriginal peoples (s13). These exceptions were carried forward in subsequent legislation.

Native and imported game came under different sets of rules within colonial legislation. For instance, native ducks were included in the *Birds Protection Act 1881* under the general category 'wild ducks of any species' within the broader category of 'native game', as opposed to 'imported game' such as pheasants (Schedules). In this period (and also throughout the twentieth century), 'wild' and 'native' were used almost interchangeably in relation to ducks, and referred to undomesticated ducks that were also in Australia before British colonial settlement in 1788 (Chew and Hamilton, 2011: 37). While both native and imported game birds were protected under colonial laws, for native birds this was to 'prevent destruction' by enforcing a closed season during the breeding season and for imported birds so that they could establish and maintain populations for hunting as sport and food (*Birds Protection Act 1881*, Preamble; see also Mackenzie, 1988: 25–53; Bonyhady, 2000: 14–39; Dow, 2008: 156–160). These measures to protect over-hunting did not always work, and there is strong evidence that in some regions of Victoria in the nineteenth century, hunters significantly reduced the populations of many native waterbirds (Dow, 2008: 156–160). In addition to open seasons, towards the end of the nineteenth century, there was also some provision for protecting birds by conserving nesting areas, a move which was linked to hunting interests. These 'game reserves' were included in the *Birds Protection Act 1881* and subsequent similar Acts. The reserves could be declared for any of the listed birds and in the 20-year period following the introduction of this Act a series of wetlands were listed as reserves (s11; Jarman and Brock, 2004: 3; Hogendyk, 2007: 14).

From the end of the nineteenth century and into the early twentieth century, arguments for the protection of birds in Australia broadened to include their value to farming, in addition to hunting interests, a shift that was reflected in the protection of insectivorous birds in various NSW state Acts (Walker, 1991: 20; Jarman and Brock, 2004: 2–6). Peter Jarman and Margaret Brock have argued that this change was partly due to intense drought in the late 1880s and 1890s, which brought the economic and environmental effects of intense over-stocking of sheep and widespread tree clearing by graziers into focus for both governments and farmers, and with it a realisation that pastoral land use practices needed to change, partly because 'native biota were suffering' (2004: 6). R. B. Walker has argued that the now largely native-born Anglo-Celtic population had a better appreciation of native flora and fauna than previous generations of mostly European migrants (1991: 19). With these motivations, throughout this period some sectors of government and advocates of conservation looked to scientists for information on 'the economic relationship that exists between animals, birds, and agriculture' (B. Nicholls, 1925, quoted in Jarman and Brock, 2004: 7), partly to counter accusations by opponents of conservation that animal and bird protection was merely sentimental. Arguments for the economic value of birds were arguments for their protection on these grounds. Publications of scientific research into the economic role of birds in agriculture, for instance, in eating invertebrates, proliferated in the 1920s and 1930s, and was a topic Kinghorn wrote about beyond his work on ducks in rice fields (1929: 263–271; Jarman and Brock, 2004: 7). 'Sportsmen' were seen to be allies in bird protection by people like Kinghorn, as they had 'long realised the necessity of the protection of game during the breeding season' (1929: 263). As agriculture intensified and expanded in the early twentieth century, particularly with new irrigation networks and closer land settlement, including from soldier settlement schemes following World War I, problems between farmers and native birds and other fauna intensified (Jarman and Brock, 2004: 6–7).

Conflicting interests and the expansion of rice growing: 1930s to 1950s

Between 1929 and 1939, approximately 20,000 to 23,500 acres were sown to rice each year in the MIA, making rice a fixture of farming there. Farmers' concerns over duck damage continued and 'special open seasons' were declared in rice areas each year from 1933, lasting from 1 September to 31 December, until 1938 (*Argus*, 1 July 1933: 27; *Sydney Morning Herald (SMH)*, 5 December 1936: 14; Ellis, 1940: 203). In that year, ducks were 'scarce' due to a prolonged dry period and state-wide open seasons were cancelled in both Victoria and NSW (*SMH*, 7 December 1937: 12; *Canberra Times*, 18 December 1937: 3; *Argus*, 8 December 1937: 17). Despite these general closed seasons, the NSW Chief Secretary declared a special open season on ducks in rice areas from 1 September (*Canberra Times*, 18 December 1937: 3). Soon after the season began, several 'natural history societies' led a protest to the NSW Government against the season having been declared at all because the months for the open season coincided with the duck breeding season. Representing the protestors to the *Argus*, the Secretary of the Royal Zoological Society (based in Sydney) argued that the Chief Secretary's decision was 'high-handed and contrary to all humane ideas' (*Argus*, 8 November 1938: 3).

The protest generated a string of letters to the *SMH* by farmers, zoologists, and ornithologists. The debates remained focused on whether ducks should be hunted through their breeding season. In them, a division is evident between those in government agriculture who argued for the elimination of ducks in the MIA (and whose views were reiterated by some farmers), and zoologists and naturalists, who were becoming increasingly aligned with species protection, and not just on economic grounds. In these debates, the views of some farmers about having ultimate control over what did and did not belong on their farms, and eliminating what did not 'fit', were articulated most directly. For instance, one farmer wrote, that: 'I quite agree with the agricultural instructor who said that "all ducks should be destroyed", meaning, of course, on the Murrumbidgee area' (9 December 1938: 3). In contrast, A. Basset Hull, the President of the Royal Zoological Society, repeated the Society's 'view that the opening of the duck season on 1 September was wrong' and also unnecessary as Kinghorn's report had shown that 'ducks are not a serious pest of rice crops' (29 November 1938: 7). This kind of interest by natural history societies in the conservation of native species was relatively new in Australia, and only started in a sustained way in the twentieth century (Jarman and Brock, 2004: 5; Hutchings, 2012: 79). It is interesting to note too that in 1938 biologists made these arguments for the 'humane' or ethical protection of ducks under the banner of the societies rather than in their government roles, which suggests that the societies provided an avenue for these arguments and interests that their government work did not.[5]

A member of the Royal Ornithologists' Union raised the potential consequences of these open seasons on graziers, arguing that while the concerns of rice growers over damage from ducks 'cannot be passed over lightly, as it represents big money', ducks needed to be protected for the benefit of wool-growers who represented 'bigger money' (*SMH*, 14 December 1938: 12). Ducks were helpful to graziers as they ate grasshoppers and snails that carried liver-fluke, which could kill sheep (*Argus*, 8 December 1937: 17). The concerns of graziers and ornithologists show one of the many issues raised by the mobility of birds, as their treatment as agricultural pests in one area possibly undermined their ability to be agriculturally valuable in another. They also draw attention to escalating disputes between established graziers and newer irrigation industries as irrigated agriculture expanded.

The outcome of the protest by the natural history societies and the ensuing debates seems to have been that no special open season was declared in 1939, although the 1938 season continued. A farmer from MIA lobbied biologists in 1940 to undertake further research into the issue, arguing that some farmers continued to be concerned that ducks damaged rice crops, and others to argue that they helped the crop (Ellis, 1940: 200). However, no research, or open season, followed. This may have been due to the view of at least some biologists, evident in the debates between famers and natural history societies in 1938, that the matter had been resolved with Kinghorn's report (Editor's notes, in Ellis, 1940: 202). It may also have been due to the outbreak of war.

World War II had significant implications for rice growing in the MIA. From 1942, rice cultivation expanded to supply food aid to parts of Asia ('Rice Statistics. 1938–57' in Agriculture-Division of Marketing and Economics Correspondence 1923–73, 384B; 'Agriculture', 384B, SRNSW; Anon, 'Rice Growing in Papua and New Guinea', 1953: 294; Scott, 1985: 275–281). Both the number of farms allowed to grow rice and the limit on the acreage already under rice cultivation on each farm were increased temporarily. The state government also expanded rice cultivation to the Murray valley irrigation districts (Anon, *Rice*, 1965: 7; Haig-Muir, 1996: 66–69). This expansion occurred despite a severe drought in 1944 and 1945. Ducks were again attracted to rice fields during these dry years and the Federal Government gave some ammunition to farmers to shoot ducks. Ducks were also trapped by the mostly Italian prisoners-of-war who farmed the fields (*Townsville Bulletin*, 26 October 1942: 2; *Camperdown Chronicle*, 13 March 1945: 4).

Following the end of World War II, Australia's food aid to parts of Asia continued and, together with national food security, became a rationale behind further increasing food production. Temporary rice fields became fixed and cultivation steadily expanded, in part, sustained through soldier settlement (see O'Gorman, 2013: 107–108). The expansion of the area under irrigation was supported by new dams, particularly those built for the Snowy Mountains Hydro-Electric Scheme, which began construction in 1949. In this post-war era, many local Wiradjuri returned to the MIA farming areas as labourers for seasonal fruit picking and factory work, re-settling parts of the towns, which had been established to service the irrigation industries, particularly Griffith (Kabaila, 1995: 133–134).

During the war years, the NSW Government had continued to declare special open seasons, but on an occasional basis, for limited periods of time, and only for some areas (see *Canberra Times*, 3 February 1949: 5; *SMH*, 22 February 1952: 5). This changed in 1952 when farmers attributed a 'rice shortage' to duck damage. Following a visit to the rice areas, the Chief Secretary of NSW declared a special open season that started on 1 September 1952 and lasted for five months.[6] At the same time the Irrigation Research and Extension Committee, on behalf of the Ricegrowers' Association of Australia, asked the CSIRO to investigate the 'bionomics of wild ducks in the irrigation areas'. In 1952, Harry Frith began this research (Frith, 1957a: 32).

Ducks and the expansion of rice: Frith

This was Frith's first assignment in the Wildlife Survey Section of CSIRO, headed by Francis Ratcliffe since its creation in 1949.[7] Since 1946, Frith had been the Assistant Research Officer at an Irrigation Research Station in Griffith, a town in the MIA. His work there had concentrated on the cultivation of orange trees, but he was increasingly fascinated by the birds of the region. In 1952, he sought to transfer to the Wildlife Survey Section to pursue this interest, as the opportunity for researching ducks arose (Tyndale-Biscoe et al., 1995: 247–249; Robin, 2007a).

Frith's research was requested by the Ricegrowers' Association, which had wholly disregarded Kinghorn's study because 'his conclusions were based on two short visits to the Murrumbidgee Irrigation Area and an examination of very few ducks' (letter from Secretary Irrigation Research and Extension Committee, to Secretary CSIRO, 22 February 1952.; 'Rice damage', NAA). Frith's research was extensive, lasting a number of years: he drew on observations he had already made about the birds during his time in the region, as well as undertaking new research through to 1956, which involved gizzard analysis of 1,849 ducks and field observations (Frith, 1957a: 33–45). The species Frith concentrated on were wood duck, and 'among the river ducks, the grey teal, black duck and white-eyed duck' (ibid.: 35). Like Kinghorn, Frith found 'that a division of opinion existed among the growers themselves' (ibid.: 33).

Frith also found, like Kinghorn, that farming practices were important in limiting crop damage. He argued that the damage attributed to ducks from puddling the soils and 'other activities', could often be traced to the same farming practices described by Kinghorn, which led to there being patches where no rice germinated. In addition, he identified the problem of destruction from wind. Most significantly, Frith argued that bare patches in the crop from poor germination rates created good landing places for ducks, who could settle there and widen the opening by pushing down encircling plants. In a statement echoing that of Kinghorn's, Frith wrote, 'it was observed that even well-grown crops of rice were very rarely visited by the birds' (ibid.: 47). Both the MIA and the newer Murray Valley irrigation districts were included in Frith's study and he implied that farming practices, and thus damage, were worse in the more recent rice growing areas. He wrote that those in the MIA had been growing rice for 'more than 30 years' and in contrast in the Murray Valley, 'rice growing was begun as a wartime expedient, and was undertaken largely by graziers with no experience of the crop'; while cultivation had 'improved', Frith argued that 'the area still lags far behind the MIA' (ibid.: 34). For Frith, the consequences of farming practices on duck damage and the overall crop were his 'most significant finding' (ibid.: 49). Frith's argument was supported by the fact that special open seasons were only declared in the newer (and not the older) rice areas each year from 1953 to 1957 in response to complaints by farmers (ibid.: 32; 'NPWS Wildlife Files', SRNSW). Nevertheless, in both places, Frith found that while different ducks fed on rice at different stages in the plants' growth, the damage was not extensive and overall that 'wild ducks are not . . . a serious pest of rice crops' (1957a: 49). An exception was the wood duck, which could graze down young rice plants if not scared off the bays; 'but', Frith wrote, 'such conditions existed usually only in a neglected crop' (ibid.: 48). The broader water landscapes of the MIA had also changed, including the creation of irrigation drainage swamps, where water was directed to low-lying, ephemeral wetlands, including Fivebough, Tuckerbill, and Barren Box swamps. These now formed 'large permanent swamps' that Frith judged to be 'presumably . . . relatively sterile' from the encroachment of watergrasses, which replaced the treed vegetation as the water regime altered (ibid.: 41).

Another major aspect of Frith's work was linking the booms in duck populations with floods. Floods occurred each year from 1950 to 1952, and in 1955, allowing him to observe the birds' responses to these events (1957a: 36). The sheer numbers of ducks in the region during wet years, he suggested, led to more complaints rather than there being an increase in damage (ibid.: 48–49). While there was a general understanding among biologists and farmers that ducks followed floods, Frith's research on their responses to floods laid much of the groundwork for later research on their nomadic behaviour (*SMH*, 12 August 1939: 12; Tyndale-Biscoe *et al.*, 1995: 252).

Frith was very aware that his research would influence whether further special open seasons would be declared. He argued that the belief 'that unlimited and indiscriminate shooting in and around the irrigation areas would reduce the population of ducks or drive the birds elsewhere' was unfounded, as many of the species of ducks were highly mobile, and more ducks would appear from around eastern Australia. He also argued that hunters who visited the area during the special open seasons were more interested in 'securing ducks than protecting rice fields'. This led them to seek out ducks that were not on rice bays and then scare them onto the fields. Frith urged tighter controls on shooting and for hunters to be more like patrol groups (Frith, 1957a: 49).

Frith's views about hunting shifted across his lifetime. His father had taught him to shoot and he had hunted birds 'for the pot' since he was 8 years old. Later statements by Frith suggest that as a young man he may also have engaged in game hunting or treated subsistence hunting as a kind of sport (see below). Frith's father, a 'bushman', was also a major influence on Frith's interest in natural history, teaching him about plants and animals (Tyndale-Biscoe *et al.*, 1995: 247; Letter from H. J. Frith to Lydia Cheuang, 4 January 1965, H. J. Frith Files, AAS). Frith's ideas about hunting, and killing in general, changed during his service in Syria and Egypt in World War II. Reflecting on this in 1965, Frith wrote that because of these experiences,

I had suffered an utter revulsion towards needless and unnecessary shooting of anything. I therefore put my guns away and became, for the first time, a birdwatcher, watching birds for their own sake. I also acquired a camera and began to achieve as much fun from stalking an animal to photograph it as I had originally in stalking it to shoot it.

(Letter from H.J. Frith to Lydia Cheuang, 4 January, H. J. Frith Files, AAS)

At the same time he needed his skills as a hunter to collect birds for gizzard analysis. While Frith continued with gizzard analysis, he nevertheless appears not to have supported hunting for sport, nor to have seen 'sportsmen' as allies of conservation, as Kinghorn had done. Frith's changing views reflected broader trends in Australian, and global, ideas about conservation following World War II. His reaction to 'unnecessary shooting' was linked to a widespread ethical shift towards valuing the lives of other living things after the large number of casualties from the war (Cartmill, 1993: 204–209; Dunway, 2000; Brower, 2005). These ideas were tied into new concerns over the environmental effects of technological 'progress', which also took hold in this period, and the growing popularity of ecology and conservation science both within and beyond the sciences.

Post-war ecology and conservation science

Both Frith and Ratcliffe were part of this increasing shift towards ecology and the development of conservation science in government institutions. Ratcliffe had studied zoology at Oxford with Charles Elton under Sir Julian Huxley, and from 1929 to 1930 worked as an economic entomologist in the CSIR (the precursor of the CSIRO), researching damage from flying foxes (or fruit bats) to agriculture in Queensland (Tyndale-Biscoe et al., 1995: 250; Robin, 1998: 134–136; Robin, 1997: 69; Dunlap, 1999: 250–251; Mulligan and Hill, 2001: 182–183; Robin, 2001: 181–183; Warhurst, 2002). In 1935, Ratcliffe began work with CSIRO into widespread severe soil erosion that was devastating the pastoral industries in inland Australia. At the same time, the USA faced a similar problem with the 'Dust Bowl' of the Midwest. Robin has argued that this crisis was a turning point in government science in both countries:

The soil erosion crisis in both the USA and Australia changed the emphasis of applied science . . . [as it] could not be handled on a 'pest control' model . . . but its progressive agenda increasingly emphasized development in the long term, not instant results.

(Robin, 1997: 70)

In Australia, conservation science as it emerged in the 1930s and 1940s 'became the next important umbrella for ecological work' (Robin, 1997: 70). Although ecology had a long history in science, with its roots in the nineteenth century, it emerged as a professional discipline only in the post-war period, with strong ties to both zoology and the methods of field naturalists (Winterhalder, 1993: 18–20; Sinclair, 2001: 175–176). Ecology placed a central emphasis on the relationships between organisms and their broader 'community' (Robin, 2007b: 159).

Within this context, the Wildlife Survey Section, led by Ratcliffe, was established to undertake a biological survey of Australia. However, the resources of the department were often claimed by the need for applied research and Ratcliffe encouraged staff to incorporate ecology and conversation science into this research. Indeed, a combination of ecology and agricultural economics became a central concept in Ratcliffe's management of the Wildlife Survey Section (Robin, 1997: 70–71; Dunlap, 1999: 249–262; Mulligan and Hill, 2001: 182–183; Robin, 2007b: 161–163).

Frith's research on rice fields sought to bring together the 'ecology and economics of wild ducks', reflecting this philosophy (Frith, 1957a: 33). The ecological approach Frith took also meant that he followed the ducks off the farm, so to speak, and he dedicated another paper to their breeding, food habits and other aspects of behaviour (Frith, 1957b). From his work on ducks, Frith argued that biologists needed to better understand ducks themselves in order to better know how to act in their long-term interest (ibid.: 19; see also Frith, 1967; Frith, 1973). Later Frith wrote:

> There are two main sorts of wildlife problems; those where the animal is a problem to man and those where man is a problem to the animal. The second is by far the more common and nearly always the more important.
>
> *(unpublished book manuscript, H. J. Frith, Files, AAS)*

The intensification and expansion of agriculture after World War II raised concerns among biologists and more widely about reductions in habitat for native animals and birds. While new dams and irrigated farms were largely seen as central to national development, they also 'set the context for the expansion of [protected] parks and reserves that characterised conservation in this period' (Jarman and Brock, 2004: 7). This was reflected in the NSW *Fauna Protection Act 1948*, which placed a new emphasis on protecting habitat. The faunal reserves created under this Act simultaneously created recreation areas for an increasingly urbanised population. The Act established a Fauna Protection Panel and brought together previously disparate laws about native birds and mammals, including pest control and wildlife protection. In this and subsequent Acts, including the 1967 and 1974 *National Parks and Wildlife Acts*, through which the NSW National Parks and Wildlife service was created, birds and animals that were seen as agricultural pests were listed as unprotected (Stubbs, 2001: 46; Jarman and Brock, 2004: 8–9). Ducks were, and are, protected under state wildlife laws, except during declared open seasons. From 1995, a hold was placed on state-wide open seasons on ducks, effectively banning this shooting sport in NSW. This ban responded to increasing concerns among many groups about the ethics of hunting for sport (Kingsford *et al.*, 2000: 12; see also Dickson *et al.*, 2009).

From the time of Frith's report, special open seasons in rice areas continued to be declared most years for several months across the Australian summer, as they still are.[8] As most farmers have remained convinced that ducks damage crops, intermittent research by government ecologists into this issue has continued. In 1997, a report from one of these investigations, by Alison Curtin and Richard Kingsford, indicated that there was 'some belief' among rice farmers that previous government research, including Kinghorn's and Frith's, had been undertaken when duck damage was not at its worst and so presented an inaccurate view of the potential severity (Curtin and Kingsford, 1997: 18). Further, in the 1970s, the accuracy of traditional gizzard analysis was shown to be suspect by biologists, throwing this aspect of Kinghorn and Frith's research into doubt. Yet the 1997 report also showed a more complex picture than that presented by some farmers, suggesting, as other investigations had done, that farming methods may be a factor in attracting ducks to rice fields (ibid.: 16–17).

An emphasis on habitat protection has continued in Australian conservation efforts, in part through international pressure. In the 1970s Australia signed a range of international agreements for habitat conservation, including in 1971 the Convention on Wetlands of International Importance, or the Ramsar Convention on Wetlands (www.ramsar.org) (Jarman and Brock, 2004: 11–12). This agreement aimed to protect the loss of wetlands, and consequently biodiversity, around the world, primarily for the conservation of waterbirds. Some wetlands in Australia have been included as Ramsar sites (including Fivebough and Tuckerbill swamps in the MIA), and their management for the conservation of birds has required the government to gain the cooperation of private land owners, those whose properties cover sections of or adjoin the wetlands, including some rice farmers.

During the late twentieth century there was a shift in landscape ecology to understanding landscapes as mosaics (see, for example, Wilson, 1995: xiii). This approach takes a holistic view that includes agricultural areas as habitats. Reflecting this, in recent years, rice farmers have started to make more space for birds, animals, and plants, treating rice farms as kinds of wetlands. It is unclear whether this change is motivated by ecological/biodiversity goals or political agendas. Rice farming has come under increasing scrutiny in recent years for its traditionally high water use farming methods, the catalyst for which was an intense drought across many parts of eastern Australia that lasted in some regions from 2000 to 2010. During this drought and afterwards, the aquatic habitat created by the fields has been used by the Ricegrowers' Association of Australia to help justify the crop and has publicly made available brochures about biodiversity on rice farms (Ricegrowers' Association). There is at least an element of strategy by rice growers in presenting their farms this way during a period of heightened criticism, to show that the water used on the farms is used for more than one purpose. These efforts can be seen as a new negotiation by rice farmers for sharing these water landscapes. However, ducks have not been included in these negotiations. In their 1997 report, Curtin and Kingsford suggested that duck 'decoy feeding' (1997: 33) areas could help to attract ducks away from rice crops. These are essentially dams with vegetation that would appeal to ducks, much like the 'wildlife-friendly' farm dams that have been encouraged by a range of organisations and by the NSW Government over the last decade and a half (see, for example, Rawton, 1999). While these do not seem to have been widely taken up by farmers, perhaps because of water shortages during the drought, in mid-2012 there were signs of 'wildlife-friendly' dams that ducks were clearly enjoying. There are a range of problems with these dams, such as the dams being on private land and subject to private interests. Further, ducks may still go to rice fields as well as the dams. However, these dams also represent more hopeful possibilities for sharing these water landscapes, and ones that perhaps both Kinghorn and Frith would have liked.

Conclusion

In this region, ducks have occupied a space at the interface of wildlife and agriculture. The multiple, sometimes conflicting, roles in which ducks have been cast reflect some of the complexity at this interface: as pests, game, economically beneficial, and protected native wildlife. Each role carries a particular legacy and set of interests about how ducks can and cannot be accommodated in these land-scapes, and all need to be considered together. Kinghorn and Frith have provided focal points in tracing these controversies. Their work and professional contexts reveal some of the wider issues that have been at stake in these controversies, changing ideas about pests and conservation at this interface, and the shifting approaches in government biological sciences, from zoology and the economic conservation of species, to ecology and conservation science. All of these relation-ships and interests have been, and continue to be, negotiated within a changed and changing water landscape.

Perhaps one of the most striking features of this case is that there has never been a definite consensus among farmers, or between farmers and biologists about whether ducks significantly damage crops. These ongoing controversies show some of the inherited relationships between particular crops, farmers, and ducks. They perhaps also reflect the persistence of different ideas about what has constituted acceptable damage to a rice crop and the continuation of the view by some farmers that ducks should be eliminated from rice growing areas. In the case of Kinghorn and Frith, some farmers may have seen their recommendations as 'top-down' science, which reflected only a portion of the views of the farming community, so that Kinghorn and Frith may have unwittingly entered into regional politics. In recent years rice farmers have started to accommodate water birds in various ways, which connects with growing scientific and popular concerns about the loss of wetlands from agricultural water use as well as with ideas about the ecological relationships between farms and what lies beyond. In many ways this ability to think about the interconnections between watery landscapes in Australia is a legacy of work undertaken by Kinghorn, Frith and their contemporaries.

Acknowledgements

This chapter has greatly benefited from the thoughts, questions, visions, and assistance of many people. Jodi Frawley and Iain McCalman's symposium on 'Rethinking Invasion Ecologies' prompted me to get moving on this research, and the work of the scholars they brought together proved to be important inspiration for questions and themes that guided the chapter. I am also in debt to both for their comments on my paper and later the draft chapter. I have had formative discussions about this research with friends and colleagues, some of whom have also generously read drafts. I would particularly like to thank Jenny Atchison, James Beattie, Geoff Ginn, Leah Gibbs, Lesley Head, Jan Oosthoek, Libby Robin, and the Environmental Humanities Group at UNSW. A special thanks to Thom van Dooren who discussed ducks and rice at length with me and read more than one draft. Rose Docker gave important assistance in researching the Australian Museum archives. This chapter is based on research funded through the Australian Research Council (FL0992397).

Notes

1 'Puddle' means to muddy and otherwise disturb flooded bays while paddling.
2 I discuss the history of rice growing in more depth in O'Gorman (2013: 96–115).
3 For example, between 1920 and 1964, Kinghorn wrote 22 separate reports for various State and Federal Government departments. 'Bibliography – Reports on Fauna to Government Departments by J. R. Kinghorn'. AMS402, Papers of J.K. (Roy) Kinghorn, 1920–1964, Australian Museum.
4 The total number of farms that had a rice crop in the 1931–1932 and 1932–1933 seasons is unclear. The area under rice cultivation was approximately 20,000 acres. 'Agriculture-Division of Marketing and Economics, Correspondence, 1923–73'.384B, SR NSW.
5 This is despite there being a significant amount of overlap between the membership of the Royal Zoological Society and the employees at the Australian Museum. For instance, Kinghorn was both a society member and Museum employee (Docker, 2007; Hutchings, 2012: 80).
6 For more on the 1951–1952 season, see Frith (1957a: 33); *SMH*, 26 September 1951: 2; *SMH*, 19 September 1952: 2; *The Mail* (Adelaide), 16 February 1952: 6; *Canberra Times*, 28 February 1952: 6; *SMH*, 1 September 1952: 4; 'NPWS Wildlife Files', 1961–1988., SRNSW.
7 Frith also later headed the Wildlife Survey Section of CSIRO. Frith's work on ducks was central to his later work and career (Certificate of nomination to the Australian Academy of Science. 23 July 1971, H. J. Frith Files, Basser Library. AAS; Tyndale-Biscoe *et al.*, 1995: 252).
8 For open seasons, see 'Duck Shooting NSW Rice Fields', 1957–78. PROV; 'NPWS Wildlife Files', 1961–1988. SRNSW; and Curtin and Kingsford (1997: 24). For more on contemporary regulations of duck hunting on rice farms, see Game Council (2013).

References

Archives

Agriculture-Division of Marketing and Economics Correspondence, 1923–73. 200. 10/25451. 384A, State Records NSW.

Agriculture-Division of Marketing and Economics. Correspondence, 1923–73. 200. 10/25451. 384B. State Records NSW.

Correspondence files. C.80/26. 1926. Australian Museum.

'Duck Shooting NSW Rice Fields'. 1957–78. VPRS 11559/P0001/300. Public Records Office Victoria (PROV).

'Environment Committee', Ricegrowers' Association of Australia archives, Leeton, NSW.

H. J. Frith Files, Basser Library, Australian Academy of Science (AAS).

'Integration of Animal and Crop Production'. 200. 10/35499. 68/1132. State Records, NSW.

'NPWS Wildlife Files', 1961–1988. W181. 12/12165 (B). State Records NSW.

Papers of J. K. (Roy) Kinghorn, 1920–1964, Australian Museum.

'Press Cuttings 1', Ricegrowers' Association of Australia archives, Leeton, NSW.

'Rice damage, goose, wild duck control in Australian rice fields'. C2/8/2. A9778. National Archives of Australia (NAA).

NSW Acts

Birds Protection Act 1881.

Fauna Protection Act 1948.

Game Protection Act 1866.

Newspapers

Argus (Melbourne).

Camperdown Chronicle.

Canberra Times.

Sydney Morning Herald (SMH).

The Mail (Adelaide).

Townsville Bulletin.

Other sources

Anon. 1953. Rice Growing in Papua and New Guinea. *Agricultural Gazette of New South Wales,* June, 294–297.

Anon. 1965. *Rice: The World's Main Food Crop, Royal Easter Show Pamphlet* Sydney: New South Wales Department of Agriculture.

Australian Bureau of Statistics. 2005. Agricultural State Profile, New South Wales, 2003–04 (7123.1.55.001.). Available at: http://www.abs.gov.au/Ausstats/abs@.nsf/Lookup/975 A0F9AADA04961CA2570430078D77B?opendocument (accessed March 2012).

Australian Bureau of Statistics. 2008. Water and the Murray–Darling Basin: A Statistical Profile, 2000–01 to 2005–06 (4610.0.55.007). Available at: http://www.abs.gov.au/ausstats/abs@.nsf/Latestproducts/DC0DC8AAE4ECD727CA2574A5001F803A?Open document (accessed March 2012).

Barclay, A. 2010. A Sunburned Grain. *Rice Today,* April–June, 9, 12–17.

Bonyhady, T. 2000. *The Colonial Earth*, Melbourne, University of Melbourne Press.

Brower, M. 2005. Trophy Shots: Early North American Photographs of Nonhuman Animals and the Display of Masculine Prowess. *Society and Animals*, 13, 13–31.

Cartmill, M. 1993. *A View to Death in the Morning: Hunting and Nature Through History*, Cambridge, MA, Harvard University Press.

Chew, M. K. and Hamilton, A. L. 2011. The Rise and Fall of Biotic Nativeness: A Historical Perspective. In: Richardson, D. (ed.) *Fifty Years of Invasion Biology: The Legacy of Charles Elton*, Oxford, Blackwell Publishing, pp. 35–47.

Connell, D. 2007. *Water Politics in the Murray–Darling Basin*. Leichardt, NSW, Federation Press.

Curtin, A. L. and Kingsford, R. T. 1997. *An Analysis of the Problem of Ducks on Rice in New South Wales*, Hurstville, National Parks and Wildlife Service.

Dickson, B., Hutton, J. and Adams, B. (eds) 2009. *Recreational Hunting, Conservation and Rural Livelihoods: Science and Practice*, Chichester, Wiley-Blackwell.

Docker, R. 2007. Kinghorn, James Roy (1891–1983). In: *Australian Dictionary of Biography*, National Centre of Biography, Canberra, Australian National University. Available at: http://adb.anu.edu.au/biography/kinghorn-james-roy-12742/text22985 (accessed 4 July 2012).

van Dooren, T. 2011. Invasive Species in Penguin Worlds: An Ethical Taxonomy of Killing for Conservation. *Conservation and Society*, 9, 286–298.

Dow, C. 2008. 'A Sportsman's Paradise': The Effects of Hunting on the Avifauna of the Gippsland Lakes. *Environment and History*, 14, 145–164.

Drendel, V. 2011. 10 Years of Drought: One Farmer Tells Her Story. *Crikey*, 13 July. Available at: http://blogs.crikey.com.au/rooted/2011/07/13/10-years-of-drought-one-farmer-tells-her-story/ (accessed 18 June 2013).

Dunlap, T. R. 1997. Ecology and Environmentalism in the Anglo Settler Colonies. In: Griffiths, T. and Robin, L. (eds) *Ecology and Empire: Environmental History of Settler Societies*, Melbourne, Melbourne University Press, pp. 77–86.

Dunlap, T. R. 1999. *Nature and the English Diaspora: Environment and History in the United States, Canada, Australia, and New Zealand*, Cambridge, Cambridge University Press.

Dunway, F. 2000. Hunting with the Camera: Nature Photography, Manliness, and Modern Memory, 1890–1930. *Journal of American Studies*, 34, 207–230.

Ellis, N. S. 1940. Ducks and the Rice Industry. *Emu*, 39, 200–206

Frith, H. J. 1957a. Wild Ducks and the Rice Industry in New South Wales. *Wildlife Research*, 2, 1, 32–50.

Frith, H. J. 1957b. Breeding Movements of Wild Ducks in Inland New South Wales. *Wildlife Research*, 2, 1, 19–31.

Frith, H. J. 1967. *Waterfowl in Australia*, Sydney, Angus & Robertson.

Frith, H. J. 1973. *Wildlife Conservation*, Sydney, Angus & Robertson.

Game Council. 2013. Do You Want to Hunt Ducks in NSW This Year? NSW Government. Available at: http://www.gamecouncil.nsw.gov.au/portal.asp?p=GameBirdManagementProgram (accessed 23 July 2013).

Gammage, B. 1983. The Wiradjuri War, 1838–40. *The Push*, 16, 3–17.

Haig-Muir, M. 1996. The Wakool Wartime Rice-growing Project and its Impact on Regional Development. *Australian Economic History Review*, 39, September, 56–76.

Hogendyk, G. 2007. *The Macquarie Marshes: An Ecological History*, Melbourne, Institute of Public Affairs Occasional Paper.

Humphries, P. 2007. Historical Indigenous Use of Aquatic Resources in Australia's Murray–Darling Basin, and its Implications for River Management. *Ecological Management & Restoration*, 8, 106–113.

Hutchings, P. 2012. Foundations of Australian Science, Sydney's Natural History Legacy, and the Place of the Australian Museum. In: Lunney, D., Hutchings, P. and Hochuli, D.

(eds) *The Natural History of Sydney*, Mosman, NSW: Royal Zoological Society of NSW, pp. 74–89.

Jarman, P. and Brock, M. 2004. The Evolving Intent and Coverage of Legislation to Protect Biodiversity in New South Wales. In: Hutchings, P., Lunney, D. and Dickman, C. (eds) *Threatened Species Legislation: Is It Just an Act?* Mosman, NSW, Royal Zoological Society of New South Wales, pp. 1–19.

Kabaila, P. R. 1995. *Wiradjuri Places: The Murrumbidgee River Basin with a Section on Ngunawal Country*, Jamison Centre, Black Mountain Projects.

Kinghorn, J. R. 1929. Bird Protection in Australia. *Emu*, 28, 263–271.

Kinghorn, J. R. 1932. Wild Ducks Are Not a Serious Pest of Rice Crops. *Agricultural Gazette of New South Wales*, 1 August, 603–608.

Kingsford, R. 2000. Review: Ecological Impacts of Dams, Water Diversions and River Management on Floodplain Wetlands in Australia. *Australian Ecology*, 25, 109–127.

Kingsford, R. and Norman, I. 2002. Australian Waterbirds: Products of the Continent's Ecology. *Emu*, 102, 1–23.

Kingsford, R. and Thomas, R. 2004. Destruction of Wetlands and Waterbird Populations by Dams and Irrigation on the Murrumbidgee River in Arid Australia. *Environmental Management*, 34, 3, 383–396.

Kingsford, R., Webb, G. and Fullagar, P. 2000. *Scientific Panel Review of Open Seasons for Waterfowl in New South Wales*, NSW National Parks and Wildlife Service, November, 4.

Kohlstedt, S. G. 1983. Australian Museums of Natural History: Public Priorities and Scientific Initiatives in the 19th Century. *Historical Records of Australian Science*, 5, 1–29.

Langford-Smith, T. and Rutherford, J. 1966. *Water and Land: Two Case Studies in Irrigation*, Canberra, Australian National University Press.

Lewis, G. 1994. *An Illustrated History of the Riverina Rice Industry*, Leeton, NSW, Ricegrowers' Co-operative Limited.

Lloyd, C. J. 1988. *Either Drought or Plenty: Water Development and Management in New South Wales*, Parramatta, NSW, Department of Water Resources New South Wales.

Mackenzie, J. M. 1988. *The Empire of Nature*, Manchester, Manchester University Press.

Main, G. 2005. *Heartland: The Regeneration of Rural Place*, Sydney, UNSW Press.

McKeown, K. 1923. List of Birds of the Murrumbidgee Irrigation Areas. *Emu*, 23, 42–48.

Mulligan, M. and Hill, S. 2001. *Ecological Pioneers: A Social History of Australian Thought and Action*, Cambridge, Cambridge University Press.

New South Wales Department of Agriculture c.1948. *Growing Rice in New South Wales*, Sydney, New South Wales Government Printer.

O'Gorman, E. 2012. *Flood Country: An Environmental History of the Murray-Darling Basin*, Collingwood, Vic., CSIRO Publishing.

O'Gorman, E. 2013. Growing Rice on the Murrumbidgee River: Cultures, Politics, and Practices of Food Production and Water Use. 1900 to 2012, *Journal of Australian Studies*, 37, 96–115.

Pittock, B., Abbs, D., Suppiah, R. and Jones, R. 2006. Climatic Background to Past and Future Floods in Australia. In: Poiani A. (ed.) *Floods in an Arid Continent*, Advances in Ecological Research No. 39, California, Elsevier, pp. 17–18.

Powell, J. M. 1989. *Watering the Garden State: Water, Land and Community in Victoria, 1834–1988*, Sydney, Allen & Unwin.

Rawton, C. 1999. Making Your Dam 'Wildlife Friendly', Land for Wildlife Note No. 2. Land for Wildlife: Voluntary Wildlife Conservation, April.

Ricegrowers' Association of Australia n.d. Rice farms Promote Biodiversity. Available at: http://www.aboutrice.com/facts/fact06.html (accessed 16 June 2012).

Robin, L. 1997. Ecology: A Science of Empire? In: Griffiths, T. and Robin, L. (eds) *Ecology and Empire: Environmental History of Settler Societies*, Melbourne, Melbourne University Press, pp. 63–75.

Robin, L. 1998. *Defending the Little Desert: The Rise of Ecological Consciousness in Australia*, Carlton South, Vic., Melbourne University Press, pp. 134–136.

Robin, L. 2001. *The Flight of the Emu: A Hundred Years of Australian Ornithology 1901–2001*, Carlton, Vic., Melbourne University Press.

Robin, L. 2007a. Frith, Harold James (Harry) (1921–1982). In: *Australian Dictionary of Biography*, National Centre of Biography, Canberra, Australian National University. Available at: http://adb.anu.edu.au/biography/frith-harold-james-harry-12517/text22523 (accessed 5 July 2012).

Robin, L. 2007b. *How a Continent Created a Nation*, Sydney, UNSW Press.

Rose, D. 2004. *Reports from a Wild Country: Ethics for Decolonisation*, Sydney, UNSW Press.

Rose, D. 2006. What if the Angel of History Were a Dog? *Cultural Studies Review*, 12, 67–78.

Roshier, D. 2009. Grey Teal: Survivors in a Changing World. In: Robin, L., Heinsohn, R. and Joseph, L. (eds) *Boom & Bust: Bird Stories for a Dry Country*, Collingwood, Vic., CSIRO Publishing, pp. 75–94.

Roshier, D., Robertson, A. and Kingsford, R. 2002. Responses of Waterbirds to Flooding in an Arid Region of Australia and Implications for Conservation. *Biological Conservation*, 106, 399–411.

Scott, J. C. 1985. An Approach to the Problems of Food Supply in Southeast Asia During World War Two. In: Martin, B. and Milward, A. S. (eds) *Agriculture and Food Supply in the Second World War*, St. Katharinen, Germany, Scripta Mercaturae Verlag, pp. 269–282.

Sinclair, P. 2001. *The Murray: A River and Its People*, Melbourne, Melbourne University Press.

Stubbs, B. J. 2001. From 'Useless Brutes' to National Treasures: A Century of Evolving Attitudes towards Native Fauna in New South Wales. 1860s to 1960s, *Environment and History*, 7, 23–56.

Tyndale-Biscoe, C. H., Calaby, J. H. and Davies, S. J. J. F. 1995. Harold James Frith, 1921–1982. *Historical Records of Australian Science*, 10, 247–263.

Walker, R. B. 1991. Fauna and Flora Protection in New South Wales, 1866–1948. *Journal of Australian Studies*, 15, 17–28.

Warhurst, J. 2002. Ratcliffe, Francis Noble (1904–1970). In: *Australian Dictionary of Biography*, Canberra, National Centre of Biography, Australian National University. Available at: http://adb.anu.edu.au/biography/ratcliffe-francis-noble-11490/text20491 (accessed 19 March 2013).

Waterhouse, R. 2005. *The Vision Splendid: A Social and Cultural History of Rural Australia*, Fremantle, WA, Curtin University Press.

Watkins, W. R. 1926. Rice-growing. Its Possibilities on the Murrumbidgee Irrigation Areas. *Agricultural Gazette of NSW*, October, 741–748.

Weir, J. 2009. *Murray River Country: An Ecological Dialogue with Traditional Owners*, Canberra, Aboriginal Studies Press.

Wilson, E. O. 1995. Foreword. In: Forman, R. T. T. *Land Mosaics: The Ecology of Landscapes and Regions*, Cambridge, Cambridge University Press, pp. xiii–xiv.

Wilson, R. M. 2012. *Seeking Refuge: Birds and Landscapes of the Pacific Flyway*, Seattle, University of Washington Press.

Winterhalder, B. 1993. Concepts in Historical Ecology: The View from Evolutionary Ecology. In: Crumly, C. (ed.) *Historical Ecology: Cultural Knowledge and Changing Landscapes*. Santa Fe, NM, School of American Research Press, pp. 17–41.

15

INVASION OF THE CROCODILES

Simon Pooley

The film *Invasion of the Crocodiles*, 2007, first shown on BBC Natural World in 2007, took its title from the assertion that 'Australia's deadly saltwater crocs are making a dramatic comeback [and] are spreading in alarming numbers'. Publicity for the film stated that 'hundreds of cattle are being killed, and most worrying of all, attacks on people are increasing every year, often in places where crocs were previously unknown' (BBC, 2007). These brief statements bring up a series of issues central to the idea of ecological invasions, including the distinction between desirable and undesirable animals, and the spatial and temporal dimensions of the concept of invasions. However, in this case the desirable animals are introduced, and the undesirable ones are 'native'.

Crocodilians (crocodiles, alligators and the gharial), while they predate our species by millennia, are often represented as unwelcome intruders. In a sense, they could be regarded as such in this volume, not being 'invasive aliens' in any technical sense. In this chapter, I show that the scientific sub-discipline of invasion biology provides a useful arena for unpacking some of the cultural assumptions bundled up in assertions of ecological 'invasions'. These attempts to define invasiveness, alienness and nativeness can be used to counter misleading popular usages of the term 'invasions'.

This chapter first discusses some key definitions used by invasion ecologists. Temporal and spatial dimensions are central, as is the notion of harm. The discussion of the temporal dimension includes brief histories of crocodilians, and crocodilians and humans, in Australia. The discussion of spatial dimensions also touches on the notion of place, and Australian ideas about nativeness. The discussion of harm focuses on crocodiles as predators, and human–crocodile conflict.

Definitions

Article 8h of the Convention on Biological Diversity (Convention on Biological Diversity, 1993) requests contracting parties to: 'Prevent the introduction of, control or eradicate those alien species which threaten ecosystems, habitats or species.' As Mooney notes, 'a strictly ecological definition' of invasion refers 'only to the rate of spread' (2005: 5). If only those invasive species which are a threat (a subjective judgement) are to be targeted, then these would seem to fit the criteria for 'pests' or in the case of animals, 'vermin' – criteria which are defined without reference to origin. Indeed, so-called native species (on a continental level certainly) may prove invasive, and harmful, e.g. *Acacia ataxacantha* in South Africa's Eastern Cape, or native daphne (*Pittosporum undulatum*) in the states of Victoria and New South Wales in Australia.

While Article 8h concerns threats to 'ecosystems, habitats or species', in practice, most targeted invasive alien species threaten local economies, and the 'goods and services provided by natural systems on which society depends' (Mooney, 2005: 5). These include cloggers of waterworks, destroyers of grazing land, and forests, stimulators of fire, crop decimators, promoters of animal diseases, and so forth (ibid.: 6).

The Convention's definition was subsequently expanded, as outlined in McNeely's IUCN collection on 'human dimensions of invasive alien species' (see Box 15.1).

These definitions are still not straightforward. Some key aspects to be unpacked include what timescale will be used to assess 'normal past or present distribution', how experts judge which species are economically and environmentally harmful

---

BOX 15.1 DEFINITIONS OF ALIEN, INVASIVE AND NATIVE SPECIES

'Alien species' are 'a species, subspecies, or lower taxon introduced outside its normal past or present distribution . . .'

'Invasive alien species' are 'an alien species whose establishment and spread threaten ecosystems, habitats or species with economic or environmental harm.'

'Native species' are 'a species, subspecies or lower taxon living within its natural range (past or present), including the area it can reach and occupy using its own legs, wings, wind/water-borne or other dispersal systems, even if it is seldom found there.'

McNeely (2001: 3).

and to what extent this fits with popular understandings/attitudes, and of course the concept of 'natural'.

The species which have been targeted by invasion biologists are, then, a hybrid of the following: 'invasive species' which take over ecosystems, which are 'aliens' (from elsewhere), and are also 'pests' in that they are judged to cause aesthetic or economic damage to the systems they invade. It should be apparent that crocodiles only very marginally fulfil the requirements for definition as invasive species, and only (arguably) as pertains to 'harm'. It is arguable that they negatively impact on the economic activities and amenity value of systems they inhabit (they also facilitate economic activity, i.e. ecotourism and crocodile ranching). However they are not technically 'aliens' if they are recolonising habitat they inhabited in historical times. It might be argued that they 'take over ecosystems' in that they render whole waterways possibly unsafe for humans, but there is no biological sense in which this constitutes an ecological invasion.

Time/scale

Which temporal scale is appropriate for determining nativeness? On a geological scale, the fossil evidence indicates that what we know as Australia has had croco-dylians (the order Crocodylia) since the mid-cretaceous period (144–65 mya), since before it was detached from Antarctica and South America (that is, before Gondwanaland split up). Many species (some terrestrial) evolved, and crocodilians occurred from Western Australia through central Northern Territory, the coasts and interiors of Queensland and New South Wales, and in the interior of South Australia (Willis, 1991: 291). We know from modern crocodiles that they inhabit areas where the average temperature of the year's coldest month remains above 10–15°C (50–59°F), though alligators can survive temperatures as low as 4°C (39.2°F) (Sues, 1992: 24–25). In past interglacial times, then, the Crocodylia were much more widespread than they are today.

The ancestors of all living crocodylians were the Eusuchia ('true crocodiles') and the first Eusuchian found in Australia is *Isisfordia duncani*, an extinct genus related to the Crocodylia from the mid-Cretaceous. Named after the Australian town Isisford, near which it was found, and the town's mayor who found it, it was the basis of claims that the Eusuchia may have originated in Gondwana (the southern supercontinent), not necessarily Laurasia (the northern supercontinent). The taxonomy is complex and details are disputed, but it seems reasonably clear that the mekosuchines were a subfamily of Crocodylians which radiated in the region during the Eocene (57.8–36.6 mya). The two found in Australia (*Pallimnarchus* and *Quinkana*) originated in the Miocene (23.7–5.3 mya), and became extinct in the Pleistocene (1.6 mya–10,000 years ago). They were replaced by the Indopacific *Crocodylus*, whose surviving members include *C. porosus and C. johnstoni* (Brochu, 2003; Salisbury et al., 2006).[1]

Neither of the two species now existent in Australia is descended from the ancient species – saltwater or estuarine crocodiles (*Crocodylus porosus*) are thought to have migrated from Africa or Southeast Asia within the past 5 million years (early Pliocene), and the freshwater (*C. johnstoni*) since the Pleistocene. Both species predate humans, as a species, by hundreds of thousands of years.

The point is that our decisions on time scales for nativeness are relatively arbitrary, given how the range of selection alters what qualifies as native or introduced. If we consider the detachment of Australia from Antarctica as 'ground zero', then arguably all of the indigenous crocodilians are extinct. If we consider the period before hominids existed, i.e. early Pliocene and before, then saltwater crocodiles may qualify. If it is to be before humans arrived in Australia, then both of Australia's current species of crocodiles are comfortably included. In practice, both species are regarded as native by natural scientists.

Humans and crocodiles in history

In Australia, crocodiles have been regarded as both sacred and as natural resources by Aboriginal peoples from the earliest period of human settlement. European settlers largely ignored them until cattle became established in northern Australia from the late nineteenth century, after which they were regarded as vermin. Crocodile populations were decimated following the Second World War, in response to high demand for skins by the fashion and leather goods industries. Following protection in the 1960s and 1970s, populations rebounded, and from the 1980s an incentive-driven conservation strategy has been followed, encouraging sustainable harvesting of wild crocodiles and their eggs. 'Problem animals' are removed or killed.

Freshwater crocodiles appear in the well-known Aboriginal rock art of the Arnhem Land plateau from around 20,000 years ago. Saltwater crocs appear in this art following the post-glacial rise in sea levels which ushered in the 'estuarine period', from c. 8,000 years ago (Department of Arts and Museums, Northern Territory, 2013). Crocodiles feature in the dreamtime stories and songlines of several northern Australian Aboriginal clans and continue to be of totemic significance to some clans today (for instance, the Gunwinggu in the Alligator rivers region and the Gumatj and Madarrpa clans of the Yolngu in East Arnhem Land). Some believe that the spirits of the dead are contained in the bodies of certain large crocodiles and the deaths of such crocodiles cause widespread mourning.

The British colonised Australia in 1788, and by the mid-nineteenth century a conception of a settler Australian identity was emerging that connected urban Australians with the bush. This was later reflected in the literary works of A. B. 'Banjo' Paterson, Henry Lawson and others. What is striking about the poems and novels which feature the Australian bush, at least to an interested alien like myself, is the absence of crocodiles. A rare exception in the Arts is a series of paintings by Thomas Baines in the mid-1850s, notably the striking painting 'Baines and Humphrey killing an alligator on the . . . Victoria River' (Figure 15.1) (1857).[2] In the exploration literature, there are a few mentions of crocodiles as encountered on naval explorations of the rivers, but even here we seldom read of notable concentrations of crocodiles. A scan of Australian newspapers to 1900 reveals an

FIGURE 15.1   Baines and Humphrey killing an alligator on the Horse Shoe Flats, near Curiosity Peak, Victoria River (Thomas Baines, 1857).

Source: Collection of the Royal Geographic Society.

intermittent interest in crocodiles, but curiously most stories concern crocodiles elsewhere in the British Empire – particularly in Africa, India and (then) Ceylon. It is after more determined efforts were made to establish a cattle industry in the late 1800s that saltwater crocodiles became objects of greater notice – and were routinely shot as vermin (Webb and Manolis, 1998: 131, 132).

The wetlands of the coastal regions of Western Australia, Northern Territory, and western Queensland, remain little developed, and together with the arid interior, form part of the mythology of the Outback. The myth of wilderness persists for this region. Alongside and imbricated with it is the equally powerful myth that Aboriginal peoples have always trodden lightly on the earth and have had no significant deleterious impacts on the biota since their arrival 50–45 kya. If it is true that the first humans to arrive in Australia had a hand in the extinction of the megafauna, then it may be that the mekosuchine crocodylians were among those species rendered extinct with their assistance. That said, the first time, inasfar as we are aware, that the country's surviving crocodile species have been threatened with extinction was when they were decimated by commercial hunters following the Second World War. Most of these hunters were of settler stock. They made extensive use of Aboriginal trackers and drew on their local knowledge (ibid., 1998: 134).

Freshwater crocodiles were hunted from the 1960s, and were soon decimated as they congregate in large numbers during the dry season. The wisdom of hunting this endemic and largely harmless species to extinction was soon questioned, and Western Australia banned hunting in 1962, followed by Northern Territory in 1963. Saltwater crocodiles – more dangerous, more disliked, and more commercially valuable – continued to be hunted, but some hunters feared for their resource. Saltwater crocodiles were given legal protection (initially for a ten-year period) from 1969 in Queensland, and from 1971 in the Northern Territory (ibid.: 134–135). Queensland remained the 'leaky barrel' from which both species shot in the former two states could be exported, until 1974. Saltwater crocodiles were only really protected from 1972 when Prof. Harry Messel, a physicist from the University of Sydney who became interested in saltwater crocodiles while testing satellite tracking technology, convinced the Commonwealth Minister of Customs to impose a total export-import ban on crocodile products in 1971. Two years later Queensland decided to protect all its crocodiles (Messel interview, 2012).

The resulting recovery in crocodile populations impacted on Aboriginal peoples living in the heartlands of crocodile habitat in Australia. This brought out some of the differences between the clans' relationships with crocodiles. Speaking at a meeting on crocodile conservation in the late 1980s, Wesley Lanhupuy, Opposition Spokesperson for Conservation in the Northern Territory Legislative Assembly, noted that for those who 'hold crocodiles sacred in the Northern Territory, there is a special relationship between crocodiles and people'. Nevertheless, he was of the opinion that 'the majority in the Northern Territory are willing to participate in soundly based ventures that will be successful for tourism or commercial gain in its own right'. This was partly a result of the recovery of crocodile populations since protection: 'people living on settlements and outstations are starting to cry out for help . . . to remove those animals living close to communities, and which have become dangerous' (Lanhupuy, 1987: 145).

Management would not be straightforward, however, as the Northern Land Council (NLC) did not feel able to speak on behalf of individual clans who had 'special ceremonial significance to crocodiles' (ibid.). Lanhupuy believed that encouraging someone like Galarrwuy Yunipuyngu, then chair of the NLC and whose totem was the crocodile, to farm crocodiles 'would be viewed very offensively' (ibid.: 146). The willingness of other clans to do so might also cause conflicts between clans. Yunipuyngu, also at the meeting, confirmed that 'my people' would not kill or eat crocodiles or collect their eggs. It might be possible to grant permission for others to do so, he said, but they would have to do it somewhere else (ibid.).

The profundity of some clans' relationship with crocodiles is well illustrated by the case of Bakurra Munyarryun of the Dhalinbuy community, who was killed by a crocodile while bathing in the Cato River in 1980. The community requested that the large crocodile believed responsible (well known to them) not be destroyed. When another man was killed by a crocodile near the same spot eight years later, they again requested that no crocodile be killed (Webb and Manolis, 1998: 118, 124).

It is important to recall that crocodiles are regarded with no such veneration by some clans, and that members of the Munyuku clan of the Yolngu hunt them and collect their eggs for food. Before commercial hunting was banned, they used to hunt crocodiles for skins which they traded for tobacco, meat, sugar, and so forth. However, unlike 'balanda' (white settlers), they did not just skin and discard the crocodile, but also cooked and ate the meat (ibid.; Quammen, 2003: 193–194). To date, traditional use of crocodiles by Aboriginal people is provided for under Section 122 of the Territory Parks and Wildlife Conservation Act, and they are not bound by hunting regulations or seasons. This right was confirmed by a decision of the High Court (Yanner vs Eaton) when Murrandoo Yanner's right to spear two crocodiles in the Gulf of Carpentaria off Queensland, and share the meat with his clan, was confirmed in 1999. This kind of traditional use is not regarded as a threat to crocodiles (Horrigan and Young, 1999; Leach *et al.*, 2009: 6, 15).

In response to a series of attacks over 12 months in 1979/80, the conservation authorities in Northern Territory responded with an 'incentive-driven conservation strategy'. After having Australia's saltwater crocodiles reclassified from the Convention on International Trade in Endangered Species (CITES) Appendix 1 (no trade) to Appendix II (controlled trade) in 1985, ranching (the collection of crocodile eggs from the wild for captive raising) was encouraged. Crocodile farmers paid landowners for permission to harvest wild eggs from their land under quota, and in 1987 the export of skins began. From 1994, landowners with crocodiles but no nesting grounds could benefit from a 'wild harvest' quota of wild crocodiles. Problem crocodiles – that is, those which have attacked or tried to attack humans, or which are judged a threat to human safety or productivity in management areas – are killed and sold for skin and meat, or captured and used as breeding stock in crocodile farms (Leach *et al.*, 2009: 3, 14).

Crocodile farming to supply the hide trade took off in Australia in the 1980s and is now a major industry in northern Australia. Both captive and wild crocodiles attract considerable national and international tourism to the region. Wild crocodiles are fed from tourist launches, yielding sensational images of big crocodiles leaping out of the water – notably the headline-grabbing image of 'Brutus' photographed on the Adelaide River in July 2011 (Dillon, 2011). Since protection, crocodile populations have rebounded, and for anyone born after the 1950s, it would seem as if crocodile numbers are unprecedentedly high. Crocodiles are dispersing in search of new living areas, moving upstream along rivers and streams into accessible freshwater habitats, and along the coast into what wildlife managers regard to be more marginal habitats.

Space and place

The spatial dimension is central to the concept of ecological invasions, yet inextricably bound up with time. To be labelled an invasive alien species, a species must be judged to have arrived from a region thought sufficiently distant (or perhaps ecologically different) to qualify as a 'non-native'. The question is, how far must a species move, or across what 'natural' ecological barriers, to qualify as an illegitimate inhabitant of its new environment? I have shown that there is no scientific basis for defining (or stigmatising) crocodiles as invasive alien species in northern Australia. However, we should not discount the deep attachments humans make to places, understood as spaces invested with human sentiment, cultural associations and memories. According to David Strohmeier, Aldo Leopold, the great American conservationist, died defending a stand of trees on his smallholding from fire. Leopold well knew that the trees would regenerate from this fire, but not in his lifetime (Strohmeier, 2005: 103). We get upset about changes to what we know, and what we believe should exist in a place, and what we know is framed within the limits of a human lifespan. If the rivers of our youth were devoid of crocodiles (after they were shot out), then their return may effect a very significant and unwelcome change to the reaches and pools we were accustomed to swim in.

Ecology has taught us, however, to be wary of shifting baselines. We should not judge what the so-called 'natural' state of any space is, based on how it appeared when first encountered. We should not judge deviation from that state to be degradation, or a corruption of the 'natural'. A hard lesson for environmentalists to learn from the 1980s was how to assess environmental degradation, if according to developments in ecological theory there is seldom a stable 'natural' baseline to measure against (Worster, 1994). This is due to 'natural' fluctuations as well as human interventions, which we increasingly discover were much more widespread and profound before European explorers and settlers arrived to describe the world than was previously recognised.

The displacing of the cyclical time of systems ecology has favoured a return to a historical conception of time in ecology, where irregular natural disturbances and periodicities of natural variation regain their importance. This implies it is necessary to investigate the environmental history of an ecosystem (a place, not a generic space or system) in order to manage it sensibly. Not to discover some past pre-human (or pre-European) baseline or 'natural state', but rather to reconstruct histories of major events and shifts, and to help establish what Szabó rightly emphasises: 'historical range of variability' (Szabó, 2010: 383).

The BBC documentary's assertion that crocodiles are being found 'often in places where crocs were previously unknown' is vague as to how long crocs were unknown to inhabit such places (in recent human memory, or a couple of generations' memory, for instance?). Given the longevity of crocodiles' habitation of Australia, it is unlikely that crocodilians have not lived previously in these areas. There were probably crocodiles there before the escalation in hunting in the 1950s. Interestingly, a letter writer to Melbourne's *The Argus* newspaper in 1875 argued that crocodiles were commonly found as far west as the Dampier Archipelago (west of what is currently regarded as their 'natural' range), and confirmed that (as today) they were found as far east as the Fitzroy River at Rockhampton ('Australia Felix', Anon, 1875). Even if crocodiles were not found previously in areas they are now inhabiting, then they almost certainly were living in these places over their longer evolutionary history. And as we see the effects of climate change, their 'natural habitat' may also undergo spatial shifts.

Australian attitudes to nativeness

Head and Muir (2004) argue that 'in Australia, the relationship between the continent and the nation has facilitated a simplistic distinction between the native species and exotic invasives in public environmental debates and the national imaginary'. The date of British colonisation (1788) is used as the 'marker of profound social and ecological change' which 'fixes the temporal threshold of nativeness for many . . .' (ibid.: 202). Deciding what is indigenous and what 'belongs' is complicated, they maintain, in an ex-colony 'where settler human populations are still coming to terms with their own belonging, particularly in relation to the indigenous prior inhabitants' (ibid.: 203).

Head and Muir note (as do others) that Aboriginal peoples had been responsible for significant environmental change for thousands of years prior to 1788 (ibid.: 202). Problematically for simplistic attempts to link indigenous peoples and the country's 'native' biota, Aboriginal people have opposed the destruction of introduced species such as the water buffalo, which they now hunt and which they have incorporated into their ceremonial lives. Aboriginal people have made intellectual space for introduced animals and accepted their role in Australian ecosystems (Trigger *et al.*, 2008: 1274). Water buffalo are of particular interest in this regard because they have been very destructive of crocodile nesting habitat. From the perspective of Northern Territory wildlife management authorities, one of the environmental gains of the legalised, controlled crocodile egg harvest has been improved management of feral animals and weeds. Water buffalo, feral pigs, and introduced plants such as *Mimosa pigra* are controlled by landowners because they are judged harmful to crocodile nesting habitat (Leach *et al.*, 2009: 4, 16).

The Wilderness Society's 'Wild Country Initiative' (WCI) seeks to 'imagine Australia as it was before 1788' when Europeans arrived, using that as point zero for the onset of environmental degradation of the country. In a study of WCI programmes Trigger *et al.* (2010) found that attitudes to ecological restoration to desirable landscapes differed between southwestern Australia and northern Australia. In a survey of farmers engaged in a WCI to preserve the malleefowl, Trigger *et al.* found that in the southwest, restoration was 'conceived in terms of transforming established agricultural landscapes' (ibid.: 1070). In contrast, in the north where the landscape was regarded as still undeveloped wilderness under the stewardship of Aboriginals believed to have a more natural relationship with the land, conservation was stressed over restoration. In the southwest, the socio-economic context and the land-use culture of settler farmers generated a vision of the restoration of a species within the framework of a mosaic of working/worked agricultural landscapes and undeveloped or 'restored' areas. In the north, the landscape is envisioned as predominantly natural, and removal (of harmful species which don't 'belong', like the invasive cane toad) is the dominant response.

Turning to crocodiles in particular, it might seem that it would be a cultural disadvantage for 'saltwater crocodiles' (*C. porosus*) that they are widely distributed elsewhere (from India in the west to Vanuatu in the east), and travel up to 1,000km across the ocean. In this sense only local populations (rather than the species) are 'Australian' (another potential calibration for non/nativeness). Certainly, it is easier to make a case for Australianness for the endemic Johnstone's crocodile. In 1884, Professor Ralph Tate of Adelaide University referred in a lecture on 'Climate and distribution of fauna in Australia' (Tate, 1884: 15) to 'two species of crocodiles, one Oriental and the other peculiar to Australia'. The common name 'saltwater' implies a marginal existence, a liminal creature, an interloper from foreign parts, a vagabond of the surrounding ocean. However, both of its popular common names – the other being the 'estuarine' crocodile – are misleading. While it is the living crocodile best adapted to living in saltwater conditions, *Crocodylus porosus* is essentially a freshwater species, and populations of saltwater crocodiles have lived for generations in freshwater systems in northern Australia (Webb *et al.*, 2010).

This question of the nativeness or otherwise of saltwater crocodiles parallels perhaps the supposedly insecure claims to indigeneity of 'settlers' in the country. As 'races' rather than species, Australians of European, Asian and other origin, are not considered 'of Australia' in the same way that Aboriginal peoples are, even though some of them are the descendants of individuals who arrived two centuries ago. Crocodiles, large and dangerous creatures – in a continent bereft of large predators – actually only became wildlife icons of the Australian bush following the media successes of the *Crocodile Dundee* films (1986, 1988, 2001), and the late Steve Irwin (whose television series *The Crocodile Hunter* ran from 1997 to 2004).[3] There is some irony here because the fictional Mick 'Crocodile' Dundee character in the *Crocodile Dundee* films was mauled by a croc and his main interaction with crocodiles in the first film is to kill one which attacks the woman he's trying to impress. In the final film in the series, because crocodile hunting is banned, he is reduced to wrestling crocodiles to entertain tourists. Despite his nickname, Steve Irwin by no means focused on crocodiles. He did oppose the proposed safari hunting of crocodiles, as have his father Bill and wife Terri in recent times. This puts them (among others) at loggerheads with Northern Territory wildlife managers and some prominent crocodile conservationists who advocate safari hunting of crocodiles as part of the existing sustainable use programme in order to further incentivise protection of wild crocodiles and their habitats.[4]

Anecdotally, at least in recent decades, few anxieties over the non-nativeness of saltwater crocodiles appear to trouble the public in northern Australia. In a news roundup for the year 2008, an *NT News* journalist (Watkins, 2009) went as far as labelling crocodiles 'NT's beloved predator'. In 2012, a feature listing the '150 most powerful in NT' (Northern Territory), including politicians, business people and sports coaches, rated crocodiles at no.14 (the only animal on the list). The feature (*NT News*, 2012) noted that:

Crocodylus porosus has been good and bad for the Territory. Tourists flock here to see the reptile that walked with dinosaurs and chic women in Tokyo and Paris love carrying a bit of it over their shoulder. But there has been another death this year, reminding us all that crocs are among the most dangerous animals on earth.

Harm

Harmfulness is an essential element in the definition of invasive alien species, and the reality of harmful invasions should not be denied. For *Crocodylus johnstoni*, the spread of introduced cane toads (*Bufus marinus*) was initially disastrous, with widespread die-offs first noted in 2005. Introduced to Queensland in 1935 to prey on the (sugar) cane beetle, they spread westward, decimating reptiles that prey on them, including crocodiles (Letnic *et al.*, 2008).

In this context, invasive or not, it is well to remember that crocodiles can be dangerous to humans. Even freshwater crocodiles can and will bite humans, but this is usually in self-defence, or as a result of mistaking a swimming human for other, smaller prey. On the other hand, for large saltwater crocodiles, humans are certainly on the menu. Male saltwater crocodiles reach an average maximum size of 5m, with some reaching up to 6m in length. Data on crocodile attacks suggests that most fatal attacks have been carried out by crocodiles longer than 4m. They have good vision at night and during the day, a good sense of smell and good hearing. The design of their heads, with raised eyes, ears and nose turrets, and cryptic colouring, allows them to achieve a considerable degree of stealth when hunting. They can remain underwater for long periods, with large crocs reputedly able to remain submerged for up to 3 hours (Caldicott *et al.*, 2005: 145–146, 150).

In an analysis of crocodile attacks in Australia between 1971 and 2004, Caldicott *et al.* found 62 cases of 'definite, unprovoked attacks by saltwater crocodiles, resulting in injury or death to humans' (Manolis *et al.*, 2013 update this to 100 attacks by 2012). Attacks on researchers, rangers and farmers working with wild crocodiles were excluded, as were attacks by captive animals (Caldicott *et al.*, 2005: 146). This study found a steady increase in overall attacks, but a relative decline in fatal attacks, over the period (see Figure 15.2). The increase in attacks is perhaps explicable in light of the increase in the wild population of saltwater crocodiles in the Northern Territory – from around 5,000 in 1971 to around 75,000 non-hatchlings in 2000. What is interesting is the relative decline in fatal attacks, given that there has been a marked increase in average size of the crocodiles in the population, following protection from hunting. Attacks have occurred at all seasons of the year, with most in the wet-warm season (November to April) – see Figure 15.3. They are not closely linked to the tourist season (June to August) as most attacks have involved locals or regular visitors, and also because most visitors avoid the hottest, wettest months.

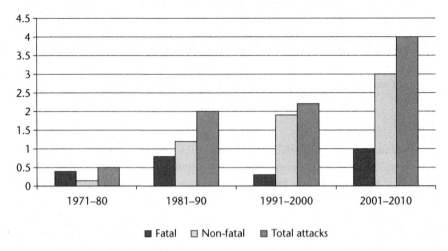

FIGURE 15.2 Saltwater crocodile attacks on humans (average per year) in Australia, 1971–2010.

Source: Adapted from Manolis *et al.* (2013).

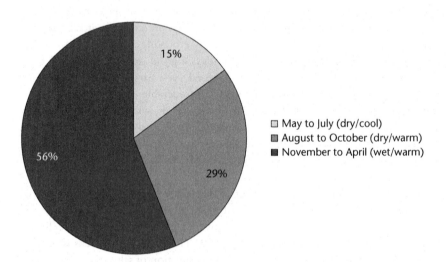

FIGURE 15.3 Seasonality of saltwater crocodile attacks in northern Australia (1971–2010).

Source: Adapted from Manolis *et al.* (2013).

Conclusion

Considering their long evolutionary and prehuman history on the continent, crocodiles cannot convincingly be described as 'aliens' in Australia, and they have seldom been so described in the historical literature. It is only on very short-term human timescales that they could be described as 'invaders' intruding into 'places where crocs were previously unknown' (BBC, 2007). In terms of harm, they do pose a threat to humans and their livestock, but their rate of increase is unlikely to threaten ecosystems or disrupt ecosystems' services. The economic harm they cause is more than balanced by the economic contributions of tourism and the sale of crocodile skins and other body parts. Whereas saltwater crocodiles attack on average four humans per year, thousands of crocodiles are killed on commercial farms annually, and there is an annual controlled harvest of up to 60,000 eggs from the wild. While the psychological effects of being attacked by a large crocodile can be devastating, and the physical wounds are often serious and require prolonged treatment, many more humans are killed by other better loved animals in the region. An *NT News* article noted that, according to coroners' records, the crocodile was the seventh most likely animal to kill you in the Northern Territory, on a list topped by horses, cows and dogs (Watkins, 2009).

Thus, it seems that the high profile often given to the danger of crocodile attacks, and the appeal of a film title like 'invasion of the crocodiles', are cultural artefacts. As the only indigenous large predator on the continent, and the only one to kill and eat humans, crocodiles are freighted with considerable symbolic baggage. In part assisted by the popularity of the *Crocodile Dundee* films, and through increased ecotourism to the region, crocodiles have become icons of the Australian bush. Unfortunately, the wild and the domesticated are becoming increasingly intertwined, and it is no longer easy to make distinctions between urban Australia and the bush 'out back'.

Harriet Ritvo has drawn our attention to the messiness of our use of the term 'wildlife', arguing: 'in a world where human environmental influence extends to the highest latitudes and the deepest seas, few animal lives remained untouched by it. At least in this sense, therefore, few can be said to be completely wild.' She gives the example of wolves translocated by air to Yellowstone National Park in the USA (Ritvo, 2011: 208). Now crocodiles have lived alongside humans for millennia, and considering the numbers of humans and crocodiles in the northern Australian states (in addition to local residents, around 1.4 million tourists a year have been visiting the Northern Territory since 2005/2006, and the state's population of saltwater crocodiles surpassed 75,000 in the mid-1990s), it should be clear that attacks on people are very much the exception. It should also be clear that crocodiles have become, and are increasingly becoming, accustomed to living alongside humans, and this will lead to accidents and, in isolated cases, deliberate attacks on humans.

We should not discount the individual agency of crocodiles. After all, the crocodile that attacks, like the one that so determinedly went after Val Plumwood in Kakadu National Park in 1985, is the exception to the rule. It may have attacked first because it felt threatened, and then in hunger – both rational explanations – but it may just have been an exceptional individual that she had the misfortune to encounter.[5] This kind of idiosyncratic behaviour is not relayed in the published scientific literature, which is about generalisations and rules. However, it peppers the fireside conversations of those who work with crocodiles. These are the night thoughts of science, to which we should pay more attention. Part of my current research on crocodilian conservation involves interviewing those who research, manage and live alongside crocodiles, in the attempt to capture examples of this kind of individual behaviour by crocodiles.

Even the hunters, with their tales of thousands of marauding man-eaters, viewed merely as animated traps, recount tales of tussles with particular 'larger-than-life' crocodile characters. A famous Australian example is that of Sweetheart, a 5.1m-long saltwater crocodile who in the mid-1970s terrorised anglers on Sweets Lookout Billabong by ripping their outboard motors off the back of their boats. Between 1974 and 1979 he bumped, bit, and even upended at least six boats. He totally ignored the humans in the boats, and those he had tipped into the water. He was eventually snared but died, probably from the exertion of his struggles to get free, and is now on display in the Museum and Art Gallery of the Northern Territory in Darwin (Webb and Manolis, 1998: 116–118). This idea of agency is important, because if all crocodiles are the same, then just as some rangers suggested after Val Plumwood was rescued, grievously bitten, all crocodiles are 'man-eaters' and should be killed. Or at least, an official should go out to the attack site and shoot any large crocodile in sight.

At the heart of the debate over ecological invasions are ideas about what belongs and what is desired (and not wanted) where, and these are informed by scientific observation and theory, and just as importantly by our lifestyle choices and our values, utilitarian and spiritual. As William Cronon (1995) observed nearly 20 years ago, we need to overcome the idea that 'nature' and 'wilderness' exist in separate realms from those we view as native habitat, reserved for us. Crocodiles will not remain in the nature reserves we have set aside for them. They appear to be colonising new areas because they were shot out across their former range by humans in the 1950s and 1960s, and they are recolonising these areas because they have been protected since the 1970s and their populations are growing.

The film *Invasion of the Crocodiles* was actually an investigation into the reasons why saltwater crocodiles have been turning up in places where they have not been seen in recent decades. It followed the work of a well-respected Australian crocodile expert, Adam Britton. In his concluding piece to camera he says:

> Crocodiles definitely have a place in the Northern Territory, but we obviously have to make sure that crocs and people are not coming into conflict with each other, and that means that there are places where we have to control them, we have to keep them from killing people. It's a balancing act, and if we don't get the balance right, then the consequences for either species could be fatal.

What would be very welcome is a companion piece examining the revolution in thinking about nature, environmental change and belonging which is required to unpack the baggage associated with what we call 'invasions' by creatures we fear, dislike or simply misunderstand.

We live in hybrid environments now, neither totally domesticated nor totally wild. Changes wrought by anthropocentric climate change, development and conservation shape the interactions of humans and the introduced and indigenous biota of ecosystems everywhere. We are still trying to work out the amplitude of environmental variability in ecosystems familiar to us, and to grasp the ecological consequences of unprecedented accelerations in anthropogenic ecological change. What is clear is that there is no indigenous or native nature 'out there' or 'out back' which does not bear our fingerprints. In our interactions with our planet's wilder spaces and their denizens we need to think beyond our remembered and idealised places.

Acknowledgements

St Antony's College, Oxford, provided institutional support during the development of my current research programme, which is now funded by an Imperial College London Junior Research Fellowship, based at Imperial College Conservation Science. My thanks to Adam Britton for comments on the text. Special thanks are due to Jodi Frawley and Iain McCalman for inviting me to this stimulating conference at the University of Sydney.

Notes

1   I am indebted to Christopher Brochu for patiently walking me through some of the complexities of crocodilian taxonomy – any oversimplifications or errors are entirely my own responsibility.
2   Search for 'Thomas Baines crocodile' at: http://trove.nla.gov.au (accessed 5 December 2013).
3   Notable crocs preceding these films include Sweetheart, discussed below, and the eponymous crocodile in Grahame Webb's novel *Numunwari* – and subsequently, crocodiles feature in the poems of Les Murray. Les Murray mentions or describes crocodiles in the poems 'Flood Plains on the Coast Facing Asia' and 'Kimberley Brief', for example.
4   For the Department's view: Leach *et al.* (2009) *Management Program for the Saltwater Crocodile*, p. 9; for the Irwins' opposition: see http://www.rspca.org.au/media-centre/media-release-archive/rspca-and-bob-irwin-slam-croc-safari-hunting.html (accessed 17 April 2013).
5   For Plumwood's account, see http://valplumwood.com/2008/03/08/being-prey/ (accessed 17 April 2013).

References

Anon. 1875. Australia Felix, Text books in geography, letter to the editor. *The Argus* (Melbourne), 16 November.

BBC. 2007. British Broadcasting Corporation, blurb for *Invasion of the Crocodiles*. Available at: http://www.bbc.co.uk/programmes/b007gv00 (accessed 19 April 2013).

Brochu, C. A. 2003. Phylogenetic approaches toward Crocodylian history. *Annual Review of Earth and Planetary Sciences*, 31, 357–397.

Caldicott, D. G. E., Croser, D., Manolis, C., Webb, G. and Britton, A. 2005. Crocodile attack in Australia: an analysis of its incidence and review of the pathology and management of crocodilian attacks in general. *Wilderness and Environmental Medicine*, 16, 143–159.

Convention on Biological Diversity. 1993. Available at: http://www.cbd.int/convention/articles/default.shtml?a=cbd-08 (accessed 19 April 2013).

Cronon, W. 1995. The trouble with wilderness. In: Cronon, W. (ed.) *Uncommon Ground: Toward Reinventing Nature*, New York, W. W. Norton, pp. 69–90.

Department of Arts and Museums, Northern Territory. 2013. http://www.artsandmuseums.nt.gov.au/museums/collection/history/chronology#.UW1L3rXqm30 (accessed 16 April 2013).

Dillon, M. 2011. Monster croc shock. *NT News*. Available at: http://www.ntnews.com.au/article/2011/07/12/246641_ntnews.html (accessed 19 April 2013).

Head, L. M. and Muir, P. 2004. Nativeness, invasiveness and nation in Australian plants. *Geographical Review*, 94, 199–217.

Horrigan, B. and Young, S. 1999. Murrandoo Yanner, his crocodiles, the High Court, and the Native Title implications. Available at: http://www.onlineopinion.com.au/view.asp?article=1081 (accessed 17 April 2013).

Lanhupuy, W. 1987. Australian Aboriginal attitudes to crocodile management. In: Webb, G. J. W., Manolis, S. C. and Whitehead, P. J. (eds) *Wildlife Management: Crocodiles and Alligators*, Sydney, Surrey Beatty and Sons, pp. 145–147.

Leach, G. J., Delaney, R. and Fukuda, Y. 2009. *Management Program for the Saltwater Crocodile in the Northern Territory of Australia, 2009–2014*, Darwin, Northern Territory Department of Natural Resources, Environment, the Arts and Sport.

Letnic, M., Webb, J. K. and Shine, R. 2008. Invasive cane toads (*Bufo marinus*) cause mass mortality of freshwater crocodiles (*Crocodylus johnstoni*) in tropical Australia. *Biological Conservation*, 141, 1773–1782.

Manolis, S. C., Webb, G. J. W. 2013. Assessment of Saltwater Crocodile (*Crocodylus porosus*) attacks in Australia (1971–2013): implications for management. In: *Crocodiles: Proceedings of the 22nd Working Meeting of the IUCN-SSC Crocodile Specialist Group*, Gland, Switzerland, IUCN.

McNeely, J. A. 2001. An introduction to human dimensions of invasive alien species. In: McNeely, J. A. (ed.) *The Great Reshuffling: Human Dimensions of Invasive Alien Species*, Gland, Switzerland, IUCN, pp. 5–20.

Mooney, H. A. 2005. Invasive alien species: the nature of the problem. In: Mooney, H. A., Mack, R. N., McNeely, J. A., Neville, L. E., Schei, P. J. and Waage, J. K. (eds) *Invasive Alien Species: A New Synthesis*, New York, Island Press, pp. 1–12.

NT News. 2012. The 150 most powerful in the NT. 20 December, Available at: http://www.ntnews.com.au/article/2012/12/20/316063_nt-business.html (accessed 17 April 2013).

Quammen, D. 2003. *Monster of God*, New York, W. W. Norton & Co.

Ritvo, H. 2011. Beasts in the jungle (or wherever). *Noble Cows and Hybrid Zebras: Essays on Animals and History*, Charlottesville, VA, University of Virginia Press, pp. 203–212.

Salisbury, S. W., Molnar, R. E. Frey, E. and Willis, P. M. A. 2006. The origin of modern crocodyliforms: new evidence from the Cretaceous of Australia. *Proceedings of the Royal Society B*, 273, 2439–2348.

Strohmeier, D. 2005. *Drift Smoke: Loss and Renewal in a Land of Fire*, Reno, University of Nevada Press.

Sues, H-D. 1992. The place of crocodilians in the living world. In: Ross, C. A. (ed.) *Crocodiles and Alligators*, Enderby, Blitz Editions, pp. 14–25.

Szabó, P. 2010. Why history matters in ecology: an interdisciplinary perspective. *Environmental Conservation* 37, 380–387.

Tate, R. 1884. Climate and distribution of life in Australia. *South Australian Weekly Chronicle*, 29 November, p. 15.

Trigger, D. S., Mulcock, J., Gaynor, A. and Toussaint, Y. 2008. Ecological restoration, cultural preferences and the negotiation of 'nativeness' in Australia. *Geoforum*, 39, 1273–1283.

Trigger, D. S., Toussaint, Y. and Mulcock, J. 2010. Ecological restoration in Australia: environmental discourses, landscape ideals, and the significance of human agency. *Society and Natural Resources*, 23, 1060–1074.

Watkins, E. 2009. The year that was . . . animals amok. *NT News*, 20 January. Available at: http://www.ntnews.com.au/article/2009/01/20/28435_ntnews.html (accessed 17 April 2013).

Webb, G. J. W. 1980. *Numunwari*, Sydney, Aurora Press.

Webb, G. J. W. and Manolis, C. 1998. *Australian Crocodiles: A Natural History*, Sydney, New Holland Australia.

Webb, G. J. W., Manolis, S. C. and Brien, M. L. 2010. Saltwater crocodile *Crocodylus porosus*. In: Manolis, S. C. and Stevenson, C. (eds) *Crocodiles: Status Survey and Conservation Action Plan*, 3rd edn, Darwin, Crocodile Specialist Group, pp. 99–113.

Willis, P. M. A. 1991. Review of fossil crocodilians from Australasia. *Australian Zoologist*, 30, 3, 287–298.

Worster, D. 1994. *Nature's Economy: A History of Ecological Ideas*, 2nd edn, Cambridge, Cambridge University Press.

Interviews

Author interview with Professor Harry Messel, Sydney, 20 June 2012.

Films

Cornell, J., director, 1988. *Crocodile Dundee* II, Paramount Pictures.

Faiman, P., director, 1986. *Crocodile Dundee*, Paramount Pictures and 20th Century Fox.

Ross, E., director, 2007. *Invasion of the Crocodiles*, Big Wave Productions Ltd., Chichester.

Wincer, S., director, 2001. *Crocodile Dundee in Los Angeles*, Paramount Pictures and Universal Studios.

INDEX

Lightning Source UK Ltd.
Milton Keynes UK
UKOW06f0409071117
312300UK00002B/141/P